Contents

Introduction

About this guide 4

AS Practical Skills 5

A2 Investigational and Practical Skills 7

■ ■ ■

Practical Skills in AS Biology

About this section 12

Practical work at AS 13

Demonstrating practical skills at AS 37

Assessment of practical tasks at AS: exemplars and comments............................. 60

■ ■ ■

Investigational and Practical Skills in A2 Biology

About this section 70

Further practical work at A2 71

Demonstrating investigational and practical skills at A2 76

Assessment of an investigation at A2: exemplars and comments 90

Answers.. 100

Introduction

About this guide

This guide is intended to help you prepare for **Unit AS 3: Assessment of Practical Skills in AS Biology** and **Unit A2 3: Assessment of Investigational and Practical Skills in Biology**. These units mainly assess 'How science works' (HSW). HSW is also assessed in the written papers, where 15% of the marks are assigned to it.

The scientific method and 'How science works'

The word science is derived from the Latin word *scientia* meaning 'knowledge'. However, science is not simply a body of knowledge. It requires an understanding of the methods by which this knowledge has been gained. Science is about how the testing of ideas allows us to develop our understanding of the natural world. You will be familiar with aspects of the scientific method from GCSE and this will be further explained in the content of this guide. Some steps in the scientific method are shown in Figure 1.

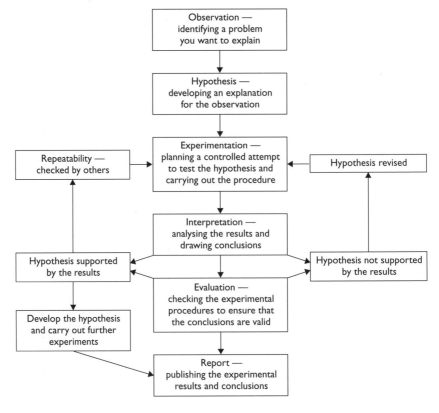

Figure 1 Steps in the scientific method

What is obvious is that science is more about *thinking* than it is about a body of facts.

'How science works' is a new component in the biology specifications. The main aspects of HSW are described below:
- Scientists use pre-existing knowledge and understanding/theories/models to suggest explanations for phenomena.
- They design, carry out, analyse and evaluate scientific investigations to test new explanations.
- They share their findings with other scientists so that they may be validated, or not, as the case may be.

As a consequence of the work of scientists, there may be implications for society as a whole. You are expected to appreciate and make informed (not emotional) comment on such aspects as:
- the ethical implications of the way in which research is carried out
- the way in which society uses science to help in decision making

How this guide is organised

This guide is divided into two parts:

1 Practical Skills in AS Biology

2 Investigational and Practical Skills in A2 Biology

Each of these parts is further sub-divided into three sections:

(a) The practical work expected to be covered within the specification;

(b) The practical skills that you are expected to demonstrate when assessed in Unit 3

(c) The marking criteria for assessment in Unit 3 and exemplar assessments with examiner's comments

AS Practical Skills

This part of the guide is divided into three sections.

Practical work at AS

This section of the guide covers all the *practical work that students are required to undertake at AS*. It includes practicals on the following:
- Biochemicals — tests for the presence of carbohydrates and proteins; and analysis of amino acids using paper chromatography
- Enzymes — procedures for controlling temperature and pH, and for producing different concentrations of solutions; experiments for investigating the effect of temperature, pH, substrate concentration and enzyme concentration; enzyme immobilisation; and following the course of an enzyme–substrate reaction using a colorimeter

- Plant cells and osmosis — measuring the water potential of a bulky tissue such as potato tuber; and measuring the solute potential of tissue at incipient plasmolysis
- Gaseous exchange — gas analysis using a J-tube; measuring oxygen consumption using a respirometer; measuring oxygen production using the Audus apparatus; and demonstrating a plant's compensation point
- Transpiration in plants — measuring water uptake using a bubble photometer
- Ecological techniques — sampling procedures (random and transect sampling); sampling devices (quadrats, pitfall traps, sweep nets and pooters); estimating abundance (density, % cover and frequency); and the influence of environmental factors on the distribution of organisms

You will be *assessed on two practical tasks* at AS. The practicals in this guide allow you to develop the techniques required for this and are likely to form the basis of the practical assessment.

In this guide each practical is outlined, but the emphasis is on the interpretation of results and the evaluation of procedures. For details of the procedures you should refer to your practical biology text, your teacher's hand-outs and your own notes on the practicals.

With each practical in this guide you are asked a question. This aims to test your understanding of the practical. Answers to these questions are provided on pages 100–01.

Demonstrating practical skills at AS

This section of the guide covers those skills that you will have to demonstrate when assessed on the practical tasks. It takes you through each of the **stages** of a practical task. At AS, these are called **skill areas**:
- Implementing a sequence of instructions
- Recording and communicating results
  - Tabulation of data
  - Graphical presentation of data
- Interpretation of the results
- Evaluation of the experimental design

This section will show you what you are expected to do in the two assessed practical tasks and how to complete each stage successfully. It examines the issues that are often raised in assessments and how these should be addressed.

In particular, it will help you to:
- gain an understanding of *each skill area*
- develop an understanding of terms such as **precision**, **accuracy**, **reliability** and **validity**
- learn to distinguish between **quantitative** and **qualitative data**, and between **measured** and **frequency data**

- learn how to *construct tables and graphs* appropriately
- recognise *anomalous results* and *assess the variation within replicated data*
- write concisely when *interpreting results*
- understand the different aspects of *evaluating a practical task*, such as identifying the limitations of equipment and procedures, and recognising both reliability and validity

At various points within this section there are examiner tips. These offer guidance on how to avoid the difficulties that often occur in examinations.

Assessment of practical tasks: exemplars and comments

This section of the guide shows the **skill areas** within which you will be assessed. For the purposes of assessment, each of these areas has *five skills*. Each skill will be marked by your teacher on a *three-point scale*: 0, 1 or 2 marks. The mark descriptors for these skills are shown in full in the specification, on pages 32–35, which is available on the biology microsite at **www.ccea.org.uk**.

Two **exemplar practical tasks** are presented. Each is undertaken by a student who has made mistakes that are often encountered by examiners. Each skill area is marked by an examiner and there are examiner comments. These are preceded by the icon ⓔ and suggest areas for improvement.

Since it is better for your understanding if you are *active* in doing something, there are several areas where you are asked to *Try this yourself.*

A2 Investigational and Practical Skills

This part of the guide is also divided into three sections. These follow on from the work at AS, which you are expected to be able to use at A2.

Further practical work at A2

This section of the guide covers all the *practical work that students are required to undertake at A2*. It includes practicals on the following:
- Populations — investigating the growth of a yeast population using a haemocytometer; and estimating the size of an animal population using a simple capture–recapture technique
- Respiration — measuring the respiratory quotient using a respirometer; and the use of redox indicators to demonstrate dehydrogenase activity in respiration
- Photosynthesis — investigating the effect of factors on the rate of photosynthesis using the Audus apparatus; paper chromatography of plant pigments; and demonstrating the role of hydrogen acceptors using a redox indicator

You will be *assessed on one investigation* at A2. These practicals, along with those that you have undertaken at AS, are likely to form the basis for the investigation.

In this guide each practical is outlined, but the emphasis is on the interpretation of results and the evaluation of procedures. For details of the procedures you should refer to your practical biology text, your teacher's hand-outs and your own notes on the practicals.

With each practical in this guide you are asked a question. This aims to test your understanding of the practical. Answers to these questions are provided on pages 101–02.

Demonstrating investigational and practical skills at A2

This section of the guide covers those skills which you will have to demonstrate when assessed on the investigation. It takes you through each of the **stages of a biological investigation**. At A2, these **skill areas** are:
- Planning the investigation
  - Developing a hypothesis using biological knowledge
  - Planning a procedure to test the hypothesis
  - Planning for statistical analysis
- Implementing a procedure
- Tabulation of data
- Statistical analysis of the data
- Interpretation of the results
- Evaluation of the practical procedures

The section will deal mostly with those aspects of investigational skills that are unique to A2 — that is, **planning** and **statistical analysis**.

In particular, it will help you to:
- understand the different *aspects of planning*: developing a hypothesis; planning a procedure; and planning for statistical analysis
- develop an understanding of terms such as **hypothesis** and **prediction**
- learn how to calculate the *mean, standard deviation, standard deviation (error) of the mean* and *confidence limits*
- learn how to assess statistically the *reliability of data* from a calculation of the standard deviation (error) of the mean and determination of confidence limits
- learn how to undertake *statistical tests* — the *t*-test and the χ^2 test
- learn how to select the appropriate *method of statistical analysis*
- use statistical techniques when *interpreting results* and *evaluating* the practical procedures

To help you develop an understanding of statistical techniques, part of this section focuses on describing and explaining the use of these techniques required at A-level.

At various points within this section there are examiner tips. These offer guidance on how to avoid the difficulties that often occur in examinations.

Assessment of a practical investigation: exemplars and comments

This section of the guide shows the **skill areas** within which you will be assessed. For the purposes of assessment, each of these areas has *five skills*. Each skill will be marked by your teacher on a *three-point scale*: 0, 1 or 2 marks. The mark descriptors for these skills are shown in full in the specification, on pages 68–72, which is available on the biology microsite at **www.ccea.org.uk**.

Two **exemplar investigations** are presented. Each is undertaken by a student who has made mistakes that are often encountered by examiners. Each skill area is marked by an examiner and there are examiner comments. These are preceded by the icon 🖉 and suggest areas for improvement.

Since it is better for your understanding if you are *active* in doing something, there are several areas where you are asked to *Try this yourself.*

Practical Skills
in AS Biology

This section is divided into three parts:

- **Practical work at AS** — this part covers all the practical work that you have to carry out at AS and includes practicals on biochemicals, enzymes, plant cells and osmosis, gaseous exchange, transpiration in plants and ecological techniques:

Biochemicals (AS 1)..page 13

Enzymes (AS 1)..page 15

Plant cells and osmosis (AS 1) .... page 23

Gaseous exchange in animals and plants (AS 2)..page 26

Transpiration in plants (AS 2)... page 30

Ecological techniques (AS 2) ....page 31

- **Demonstrating practical skills at AS** — this part describes the skills that you have to demonstrate when assessed on the practical tasks. These include:

Implementation ....page 37

Recording and communication

    Tabulation of data..page 43

    Graphical presentation of data..page 47

Interpretation .... page 54

Evaluation... page 56

- **Assessment of practical tasks: exemplars and comments** — this part shows the skill areas within which you are assessed. Two exemplar practical tasks are presented:

Practical task 1: the effect of temperature on membrane permeability
    in beetroot (assessing *Implementation*, *Tabulation of data* and
    Evaluation)..page 61

Practical task 2: investigating the water potential of potato and sweet
    potato (assessing *Graphical presentation of data* and *Interpretation*).... page 65

Practical work at AS

Biochemicals (AS 1)

Tests for the presence of carbohydrates and proteins

You should be familiar with the tests summarised in Table 1.

Table 1 Biochemical tests for carbohydrates and proteins

Test	Biochemical tested	Relative proportions of reagent(s) and test solution	Colour change if biochemical present — positive result
Iodine test	Starch	Few drops added to test solution; or sample of test solution added to dilute iodine	Straw yellow or orange (depends on concentration) to **blue-black**
Benedict's test	Reducing sugar — all monosaccharides and some disaccharides* (e.g. maltose)	Approximately equal volumes of Benedict's solution and test solution; these are then heated in a water bath	Blue to **green, yellow, orange and finally brick-red** (depending on amount of reducing sugar present)
Clinistix	Glucose (a specific test)	Clinistix strip is dipped in test solution (for 10 seconds)	Blue to a range of colours from **green to brown** (amount of glucose present estimated by comparing with standards shown on the packaging)
Biuret test	Proteins (soluble or globular)	Potassium hydroxide added to the test solution until it clears, and then a drop of copper sulfate added down the side of the tube; or Biuret reagent is used (contains both of above)	**Blue ring** at surface and then, on shaking, the solution turns **purple**; from blue to violet (or mauve or any shade of purple)

*The disaccharide sucrose is not a reducing sugar and gives a negative Benedict's test. Its presence can be detected provided it is first hydrolysed into glucose and fructose, and then tested with Benedict's solution, which will yield a positive result. Since this procedure would also work with starch, its absence must first be ascertained by testing with iodine solution.

Question 1
Which biochemical test requires heat to give a positive result?

Analysis of amino acids using paper chromatography

Paper chromatography may be used to separate the amino acids in a mixture. A small amount of solvent is put in the bottom of a chromatography tank (see Figure 2). Chromatography paper with a concentrated spot of mixed amino acids, on a line

13

above the end of the paper, is suspended in the tank so that its end dips into the solvent. The solvent moves slowly up the paper, carrying the amino acids with it. Each amino acid is carried a different distance according to its relative molecular mass and its solubility in the solvent. The paper is then treated with a reagent, such as ninhydrin, which stains the amino acids.

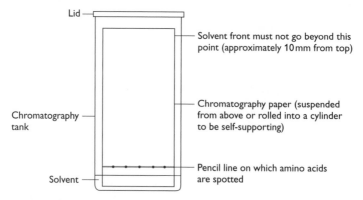

Figure 2 Paper chromatography apparatus for the analysis of amino acids

The identification of each amino acid involves the calculation of its R_f value.

$$R_f = \frac{\text{distance travelled by the amino acid}}{\text{distance travelled by the solvent front from the spotting line}}$$

Tables of R_f values are specific for the type of paper used, the solvent and the temperature. Because of this, tables of R_f values provided in many books are of little use. Known amino acids should be spotted on the same paper as the mixture. This will allow for a direct comparison of the R_f values and of the colour of the spots (which may vary from brown to purple). However, if two spots travel the same distance and have the same colour, this is not a guarantee that they are the same amino acid. Different amino acids can have very similar R_f values for any particular solvent. This problem is solved by using two-way chromatography: the amino acids are further separated by turning the paper through 90° and repeating the process at right angles to the original using a different solvent.

There are a large number of precautions to take when using paper chromatography:
- The solvent must be left in the tank (with the lid on) to create a saturated atmosphere of solvent vapour before the paper is introduced.
- The paper should only be handled sparingly and at its top edge.
- The spotting line should be marked in pencil.
- A fine pipette is used to add a spot of the amino acid mixture (to a point on the spotting line); this is dried and the process repeated to produce a concentrated spot.
- The spotting line must not be allowed to dip into the solvent.
- The solvent front must not be allowed to reach the end of the paper.
- The solvent front must be marked, in pencil, as soon as the chromatogram is removed.

- Ninhydrin is a carcinogen and so gloves must be worn, and it should only be sprayed onto the chromatogram in a fume cupboard.

Question 2

An amino acid is spotted on a line that is 30 mm from the end of the paper. After running and developing the chromatogram, the amino acid spot is found to be 84 mm from the end of the paper and the solvent front 150 mm from the end of the paper. Calculate the R_f value for this amino acid.

Enzymes (AS 1)

Properties of enzymes

You should be able to carry out experiments to investigate the following properties of enzymes:
- the effect of temperature
- the effect of pH
- the effect of substrate concentration
- the effect of enzyme concentration

Use of water baths to change or control temperature

There are two types of heated water bath that you can use.
- You can use a beaker of water heated, using a Bunsen burner, to the desired temperature, checking this with a thermometer. Since the water will cool down, the water must be heated intermittently to just above the temperature needed and then allowed to fall to just below this level. With regular attention, the temperature can be maintained to within ±2°C.
- You could use a thermostatically controlled water bath. It does not keep the temperature completely constant but is more precise than using a Bunsen burner to heat a beaker of water. The temperature should be checked with a thermometer, initially for establishing the desired temperature and then regularly thereafter to monitor any fluctuations.

A low temperature, such as 10°C, can be achieved by using a mixture of water and ice in a beaker. The temperature is monitored using a thermometer and adjusted by adding more water or ice as appropriate.

Use of buffers to change or control pH

The pH scale is a measure of the acidity or alkalinity of a solution. It is an indication of the concentration of hydrogen ions relative to water. The pH of water is 7. Solutions with a higher concentration of hydrogen ions than water have a lower pH and are acidic. Solutions with a lower hydrogen ion concentration than water have a higher pH and are alkaline. pH indicators change colour to reveal the acidity or alkalinity of solutions. The pH scale and corresponding colours of universal indicator are shown in Figure 3.

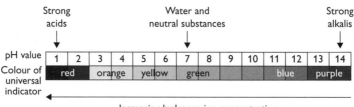

Figure 3 The pH scale

Adding just a tiny amount of a strong acid (or strong alkali) to a neutral solution changes its pH dramatically. **Buffers** are solutions that resist this change and help to maintain a constant pH. Buffers can be prepared to maintain the pH of a solution at any given value.

Use of different concentrations of solutions

To prepare different concentrations of solutions it is usual to prepare a stock solution, say a 10% solution (by dissolving 10 g of solid in 100 cm³ of water). There are two types of dilution.

- One type of dilution produces an arithmetic series of concentrations — for example, 10%, 9%, 8%, 7%. Table 2 shows how you would prepare such an arithmetic series.

Table 2 Preparing an arithmetic series of dilutions

Volume of 10% stock solution/cm³	10	9	8	7	6	5	4	3	2	1
Volume of water/cm³	0	1	2	3	4	5	6	7	8	9
Concentration of solution/%	10	9	8	7	6	5	4	3	2	1

- Another type of dilution produces a logarithmic series of concentrations. This is serial dilution. In a serial dilution, each solution along the series is less concentrated than the previous by a set factor. Figure 4 shows how you would undertake a doubling dilution so that subsequent dilutions are half the concentration of the previous solution.

Other dilution factors can be used, commonly ×10. To get a ten-fold dilution, you add 1 cm³ stock solution to 9 cm³ water, 1 cm³ of this dilution to 9 cm³ water, and so on. Serial dilution produces a wide range of dilutions — the doubling dilution, shown in Figure 4, produces a range of 10% to 0.156% solutions; the bigger the dilution factor the greater the range. You need to be careful when plotting this range on a graph.

Sometimes dilutions are made with buffer solutions so that the pH of each dilution is the same.

There are many enzyme systems that can be used to investigate the influence of pH, temperature, substrate concentration and enzyme concentration. Investigations of these are shown below, each with a different enzyme system.

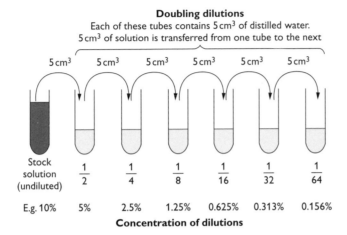

Doubling dilutions
Each of these tubes contains 5 cm³ of distilled water.
5 cm³ of solution is transferred from one tube to the next

Concentration of dilutions

Figure 4 Preparing a serial dilution — doubling dilutions

Effect of temperature on the activity of lipase

Phenolphthalein is an indicator that is pink in alkaline solutions of about pH 10. When the pH drops below pH 8.3 phenolphthalein goes colourless. In this experiment an alkaline solution of milk, lipase and phenolphthalein will change from pink to colourless as the fat in the milk is hydrolysed to produce fatty acids (and glycerol), thus reducing the pH to below 8.3.

Milk is mixed with sodium carbonate solution to make the solution alkaline and 5 drops of phenolphthalein are added. A series of water baths is set to a range of temperatures. Each water bath has placed in it a test tube containing the milk and a small beaker of lipase. When the temperature of the solutions has equilibrated to the temperature of the water bath (say, after 10 minutes) lipase is added to the test tube containing the milk and a stopwatch started. The time taken for the mixture to turn from pink to colourless is a measure of enzyme activity. The volumes and concentrations are controlled. Temperature is the variable that you are changing. The 'time taken' can be converted to a rate of reaction by calculating its inverse. In any case, a short time indicates a high rate of reaction, which you would expect to be optimal at approximately 40°C. Results are recorded in a table and plotted on a graph.

Interpretation of the results will require you to explain an increase in lipase activity up to the optimum temperature, with reference to kinetic theory (molecules moving more rapidly and so colliding more frequently) and a decrease above the optimum as a result of lipase denaturation (specific bonds holding the tertiary structure having been broken). In evaluation, you should refer to the subjective nature of determining 'when the solution becomes colourless'. Replication may be provided by including the results of other students — this, of course, compounds the subjectivity of measuring when the solutions have become colourless. It is possible to vary the concentration of the lipase to investigate the effect of enzyme concentration.

However, this experiment cannot be used to assess the effect of different pHs, since the mixture has to be initially alkaline. Indeed, the use of a dilute alkali (sodium carbonate solution) and the changes in pH during the course of the experiment might well have an adverse effect on the validity of any conclusions.

Question 3

Explain why the solution of lipase is placed in the water bath for 10 minutes before adding 1 cm³ to the milk (forming a reaction mixture).

Effect of pH on the activity of amylase

Measure the time taken for amylase to completely break down starch, by withdrawing samples of the reaction mixture at 10-second intervals and noting the time at which the solution no longer gives a blue-black colour with iodine solution (which should remain orange). Use buffers to provide solutions at different pHs. The rate of reaction can be calculated as '1 over time'.

Interpretation of the effect of pH should refer to an optimum rate of activity at, or near, a particular pH. As the pH differs from this optimum, changes in the hydrogen ion concentration alter charges on the amino acids that make up the active site of the enzyme. As a result, substrate molecules will bind less readily. At more extreme pHs, bonds that maintain the tertiary structure are broken so that the enzyme is denatured.

In the procedure it is important to add the amylase to the buffer prior to adding the starch solution. In evaluating the experiment, the main errors will relate to timing: that the stopwatch is started immediately the starch is mixed with the enzyme–buffer solution; and that there is no delay in sampling. A delay in sampling would cause the reaction time to be underestimated (and the rate to be overestimated). There are other issues worth discussing: there may be some difficulty in determining when a blue-black colour is no longer formed; and there must be a lack of precision in determining reaction time since samples are only taken every 10 seconds, rather than more frequently.

Question 4

State which variables you would need to control in this experiment.

Effect of substrate concentration on the activity of catalase

The breakdown of hydrogen peroxide to water and oxygen is catalysed by the enzyme catalase. This enzyme is present in all living tissues, as hydrogen peroxide is a toxic by-product of metabolism. Catalase activity can be investigated in the laboratory by determining the amount of oxygen produced in a period of time (say, 30 seconds). A series of different concentrations of hydrogen peroxide are prepared and added to a source of catalase, such as puréed potato.

There are different methods available for measuring oxygen production over this time period and each has its advantages:

- catalase solution is added, via a syringe, to hydrogen peroxide in a measuring cylinder and the resulting height of the foam generated is measured — see Figure 5(a)
- catalase solution is placed in a flask, hydrogen peroxide added via a burette and the oxygen produced collected by displacing water — see Figure 5(b)
- catalase solution is placed in a flask, hydrogen peroxide added via a burette and the oxygen produced collected in a gas syringe — see Figure 5(c)

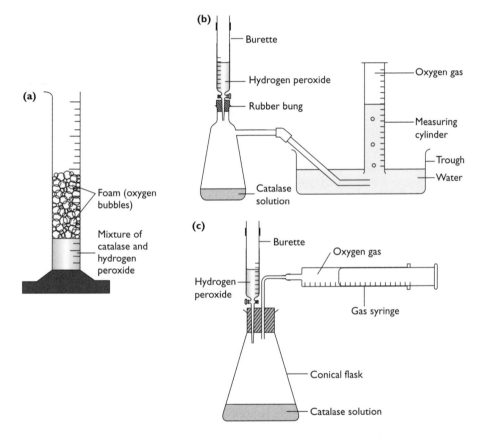

Figure 5 Measuring the amount of oxygen produced as (a) height of foam, (b) oxygen collected by displacing water and (c) oxygen collected in a gas syringe

The first method, measuring the height of the foam, is easier to set up (and can readily be repeated) but does it produce sufficiently accurate results? Some of the oxygen produced will escape; and there may be difficulty in measuring the level to which the foam rises. In the other two methods none of the oxygen can escape. Method two, collecting gas by displacing water, involves apparatus that is more difficult to set up and so taking repeat measurements will be time consuming.

Interpretation of results requires you to explain that as substrate concentration increases then collision with an enzyme molecule is more likely and so the rate of reaction increases. However, at high substrate concentration there are insufficient enzyme molecules for substrate molecules to bind to, the enzymes become saturated and so the rate of reaction reaches a maximum. In evaluation most concern will relate to the measurement of the oxygen produced: oxygen may escape, especially in the 'measurement of the height of foam' method (which makes it less suitable), and possibly from the connections in the other methods if these are not tight. In the 'displacement of water' method a little oxygen may also dissolve in the water.

Question 5

The breakdown of hydrogen peroxide is an exothermic reaction. You could detect the heat by feeling the reaction vessel. How would this heat affect the results?

Effect of enzyme concentration on the activity of protease

Exposed and developed black and white negative film consists of a plastic backing, silver salts where the film is black (exposed) and a covering of the protein gelatin. If a small square of black film is immersed in a solution containing a protease enzyme, such as pancreatic protease, this will digest the gelatin and the silver salts will be released into the solution so that the square of negative film becomes clear. The time taken for the negative to clear is a measure of enzyme activity. This set-up can be used to investigate the effect of protease concentration on its activity.

Obviously, the more enzyme that is present then the greater the likelihood of it binding with the gelatin and the faster digestion takes place. This experiment relies on determining the time for the film to clear. Not only can this be subjective but if there is a grease mark on the negative film then part of it will fail to clear.

Question 6

Explain why a piece of negative film, immersed in a solution of protease, would fail to clear if covered by a smear of oil.

Illustration of enzyme immobilisation

The enzyme sucrase can be entrapped in calcium alginate beads by following the gel immobilisation procedure. The use of the immobilised sucrase is shown in Figure 6.

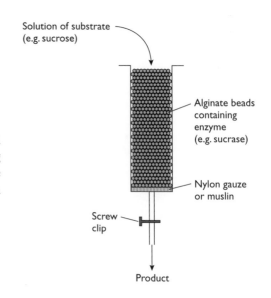

Figure 6 Apparatus used for illustrating enzyme immobilisation.

Question 7
State the names of the products of sucrose digestion. What test may be used to indicate the presence of these products?

Following the course of a starch–amylase reaction using a colorimeter

A colorimeter is set up to measure the starch content of samples taken from a reaction mixture of starch and amylase. Samples are taken at intervals (say, every 30 seconds) and added to dilute iodine. If there is any starch left it will give a blue-black colour: a dark colour means that a lot of starch is still present, a light colour means that there is comparatively little. The intensity of blue-blackness is measured using a colorimeter — see Figure 7.

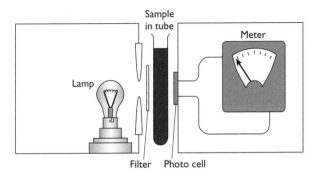

Figure 7 A colorimeter

Since the intensity of blueness is being measured, a red filter is used.

Question 8
Explain why a red filter is used in a colorimeter when measuring the 'blueness' of a solution.

Immediately before measuring the blueness of each sample, a colorimeter tube (or cuvette) containing only the dilute iodine (sometimes called the blank) is placed in the colorimeter and the instrument adjusted to 100% transmission (or zero absorbance). This standardises the instrument so that subsequent readings are comparable (this is because the intensity of the light source depends on the mains current and this can vary as other electrical appliances are switched on or off). With the use of the colorimeter you then have a series of readings (either percentage transmission or absorbance) for the samples of the reaction mixture taken at different time intervals.

These values can be plotted as % transmission (or absorbance) against time. However it would be better if the % transmission readings were converted to % starch. This is achieved by adding fixed volumes of standard solutions of starch (that is, with a known concentration, such as 1%, 0.5%, 0.25%, and so on) to dilute iodine and taking colorimeter readings. Plotting the colorimeter readings against concentration of starch gives you a curve. This is a calibration curve — an example is shown in Figure 8.

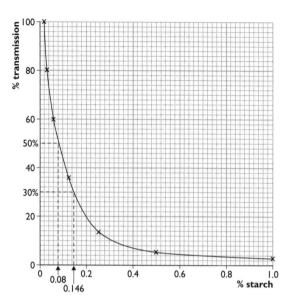

Figure 8 A calibration curve for a colorimeter: % transmission against concentration of starch

This curve is used to convert the percentage transmission readings for the starch digestion over time experiment. Two values are shown being converted in Figure 8: 30% transmission (the value for the sample taken at 1.5 minutes) converts to 0.146% starch; 50% transmission converts to 0.08% starch (the value for the sample at 2.5 minutes). Having completed this for all the results, a graph of % starch against time can be plotted. This shows the course of the starch–amylase reaction — see Figure 9.

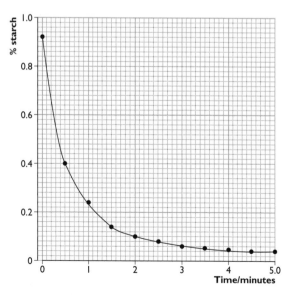

Figure 9 The course of a starch–amylase reaction

Interpretation of the trend — an exponential decline curve — requires you to explain that as starch is digested there is less left for amylase to act on. In other words, the substrate concentration decreases as the reaction progresses and so the rate of reaction is reduced over time. There are a number of important points to note in an evaluation of the experiment: firstly, the importance of the filter — a blue solution will most effectively absorb red light and this is what is transmitted through a red filter; the importance of standardising the colorimeter to 100% transmission using a 'blank' prior to taking each experimental reading; and the use of the calibration curve, since the relationship between % transmission and % starch is not linear (as shown in Figure 8). Also it is noteworthy that the value for the sample taken initially is less than 1% starch, representing a delay in sampling.

Plant cells and osmosis (AS 1)

Measuring the average water potential of cells in a plant tissue

The water potential of a plant tissue can be determined by the following principle. If a tissue shows no net gain or loss of water when immersed in a solution of known molarity, its water potential is equal to that of the external solution.

Groups of discs of potato tissue (or other suitable plant tissue) are prepared, surface-dried, and weighed — this is the initial mass. Each group of discs is immersed in one of a series of sucrose solutions of different molarity (and left for at least 30 minutes to allow osmotic changes to occur). These are removed, surface-dried (to remove *surplus* fluid) and weighed again — this is the final mass. The percentage change in mass of each group of discs (change in mass divided by the initial mass and multiplied by 100) is calculated. If the results have been repeated (for example, by other students) then mean percentage changes in mass can be calculated. This can be plotted against molarity of sucrose solution, or its water potential (using a conversion table — for example, a 1 mol dm^{-3} solution has a water potential of –3500 kPa). Figure 10 shows a graph of mean percentage change in mass against water potential of immersing sucrose solutions.

The graph shows that where the sucrose solution has a very negative

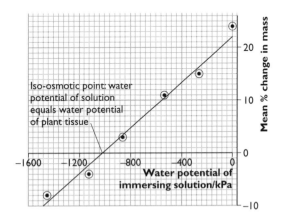

Figure 10 Mean percentage change in mass of potato tissue plotted against water potential of immersing solution to enable the determination of the water potential of potato tissue

water potential (since it has a high molarity and is more concentrated) then the tissue loses mass; this is due to water being drawn *out* osmotically and so the tissue must have a higher water potential than the immersing sucrose solution. Where the sucrose solution has a less negative water potential (is dilute) the tissue gains mass (due to water moving *in* osmotically) and so the water potential of the tissue is lower than the immersing solution. A line of best fit is drawn (using the rules provided on page 51). Where this line intersects the horizontal axis at zero percentage change in mass, you read off the water potential of the immersing sucrose solution. This is the water potential of the potato tissue. In Figure 10, this is determined as –1015 kPa.

There are important precautions to be observed in this experiment. The surface drying must be standardised: the pressure applied to the filter paper in drying the discs must be light and must be the same for each group of discs — you are only removing the *surplus* fluid. You must not squeeze the discs or they will all lose water. It is also important to work quickly to avoid loss of water from the discs through evaporation. You must be able to assume that the change in mass is due to the osmotic loss or gain of water with respect to the immersing solution.

Question 9
In the experiment described above the water potential of the immersing sucrose solution was assumed to be equal to its solute potential. Explain why this assumption is acceptable.

Measuring the average solute potential of cells at incipient plasmolysis

Incipient plasmolysis is the stage when the plant cell is just about to become plasmolysed. At this point, the pressure potential is zero because the cell wall is not exerting any pressure on the protoplast. If the pressure potential (Ψ_p) is zero, then the water potential of the cell (Ψ_{cell}) is equal to its solute potential (Ψ_s).

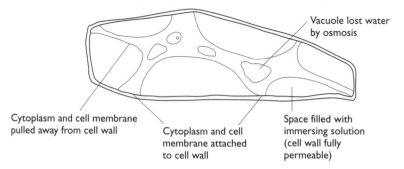

Cytoplasm and cell membrane pulled away from cell wall

Cytoplasm and cell membrane attached to cell wall

Vacuole lost water by osmosis

Space filled with immersing solution (cell wall fully permeable)

Figure 11 A plasmolysed cell

A solution with a solute potential that is equal to the solute potential of the cells will cause incipient plasmolysis. Therefore, to determine the solute potential of plant cells, you need to find the molarity of sucrose solution that causes incipient plasmolysis. Of course, it is not possible to recognise incipient plasmolysis in a cell but, since plant cells will all differ slightly, the *average* solute potential can be determined when 50% of the cells are plasmolysed. Plasmolysed cells are recognised because the cytoplasm and sap vacuole both lose water and contract, so that the cell membrane pulls away from the cell wall. A plasmolysed cell is shown in Figure 11.

In the experiment, squares of onion epidermis (or other suitable tissue) are placed in a series of sucrose solutions of different molarity, left for some time (say, 20 minutes) and then viewed using a microscope under low power. The total number of cells within the field of view and the number identified as being plasmolysed are counted. The percentage of cells plasmolysed (number plasmolysed divided by the total viewed and multiplied by 100) is plotted against molarity or solute potential of sucrose solution, and a line of best fit drawn — see Figure 12.

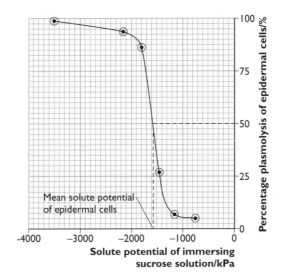

Figure 12 Percentage of epidermal cells plasmolysed plotted against solute potential of immersing solution to enable the determination of the average solute potential of the epidermal tissue

The graph shows that most cells are plasmolysed in sucrose solutions of very negative solute potential (to which they lose water osmotically); while, most cells are turgid in sucrose solutions of less negative solute potential. The average solute potential of the epidermis is determined at the point on the line at which there is 50% plasmolysis — see Figure 12.

The experiment relies on the student's ability to identify plasmolysed cells. This may be made easier by using red onion because of their coloured cell sap. It is also important that you understand the theoretical basis of the experiment: after 20 minutes, the water potential of the tissue and of the immersing solution have equilibrated (they are the same); the solution is under no pressure, so its water and solute potentials are the same; at 50% plasmolysis, the average pressure potential is zero, and so, on average, the water potential of the cells is equal to their solute potential — the solute potential of solution and tissue are the same.

Question 10

Suggest why the cells in the tissue do not all have the same solute potential. What might be the consequence of this?

Gaseous exchange in animals and plants (AS 2)

Gas analysis using a J-tube

In gas analysis, the carbon dioxide and oxygen content of air is determined. A sample of air is drawn into a J-tube (see Figure 13) using the syringe (or screw device) and the length of the air column recorded. Potassium hydroxide (a reagent that absorbs CO_2) is then drawn into the tube, shuttled backwards and forwards so that any CO_2 present in the air is absorbed, and the length of the air column re-measured. Potassium pyrogallate (which absorbs CO_2 and O_2) or pyrogallol (which will react with the KOH, forming potassium pyrogallate) is then drawn into the tube to absorb O_2 in the air, and the length of the air column measured again. The percentages of CO_2 and O_2 in the air sample are then calculated:

$$\text{percentage of } CO_2 = \frac{\text{original length} - \text{length after } CO_2 \text{ absorption}}{\text{original length}} \times 100$$

$$\text{percentage of } O_2 = \frac{\text{length after } CO_2 \text{ absorption} - \text{length after } O_2 \text{ absorption}}{\text{original length}} \times 100$$

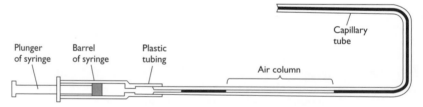

Figure 13 J-tube with a column of air drawn in for gas analysis

Gas analysis allows the comparison of the relative amounts of O_2 and CO_2 in inspired (i.e. atmospheric) and expired air, or in expired air at rest and following vigorous exercise. The technique can also be used to analyse the gas produced by an aquatic plant such as *Elodea* — it is not 100% O_2.

There are a number of precautions that need to be observed when using a J-tube:
- The J-tube must be thoroughly cleaned before use.
- Lab coats and eye protection must be worn because of the corrosive nature of the reagents.
- The J-tube must be handled only at the syringe (or screw) end to prevent warming of the tube and subsequent expansion of the air sample.

Furthermore, measuring to the nearest millimetre may lead to a lack of precision when percentages are calculated; a longer air column will reduce the error.

Question 11
Explain why potassium hydroxide treatment always precedes treatment with potassium pyrogallate (or pyrogallol).

Measuring oxygen consumption using a respirometer

Respiring organisms absorb and use O_2 from the atmosphere. However, they also produce and release CO_2. In a closed vessel the consumption of O_2 by an organism will cause a reduction in pressure if the CO_2 is chemically removed (absorbed by potassium hydroxide). The respirometer (see Figure 14) has two identical closed vessels: one contains living organisms, e.g. germinating seeds (B), and the other acts as a thermobarometer (A) — small changes in temperature or pressure cause air in this vessel to expand or contract, compensating for similar changes in the first vessel. As O_2 is consumed, the level of the fluid in the manometer will rise up in the right hand arm. The length of movement, over a set period of time, say 10 minutes, represents a measure of the rate of O_2 consumption. If the diameter (2 × radius, r) of the bore of the manometer tube is known, then the volume of O_2 consumed can be calculated (length × πr^2). Alternatively, the syringe can be depressed to introduce a volume of air needed to equalise the levels in the manometer tube and the volume read off the syringe scale.

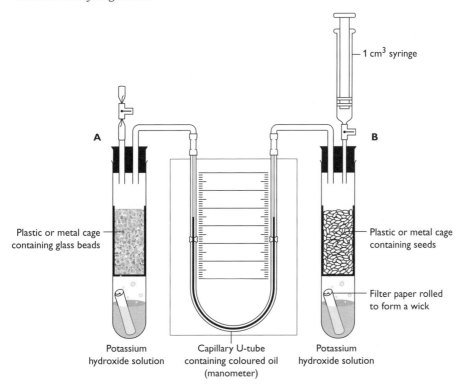

Figure 14 A respirometer for measuring the rate of oxygen uptake by small organisms

The measure of O_2 consumption of a variety of living organisms can be considered, e.g. pea seeds, mung beans, maggots, mealworms and woodlice. Weighing the mass of the organisms allows the rate of O_2 consumption to be calculated as a volume per gram of tissue per unit time ($mm^3 g^{-1} min^{-1}$).

The respirometer is very sensitive to pressure changes and needs to be assembled carefully. This can be more readily demonstrated by your teacher. The apparatus must be tested for leakages at any of the joints. A decrease in the values obtained during the course of an experiment would be due to air entering from outside. Since potassium hydroxide is corrosive, lab coats and goggles must be worn. While there are no ethical issues if using plant material, animals must be handled with care and respect.

Question 12
Explain the use of the 'wicks' in the respirometer — see Figure 14.

Measuring 'oxygen' production (by an aquatic plant) using the Audus apparatus

Aquatic plants such as Canadian pondweed (*Elodea canadensis*) produce an excess of O_2, which emerges as a stream of bubbles. It is possible to estimate O_2 production by counting the number of bubbles produced per unit time (or the time to produce a certain number of bubbles). However, since bubble size (and therefore volume) can vary, this will not yield results with any great precision. A more accurate technique involves collecting and measuring the O_2 produced. This is what a photosynthometer, such as the **Audus apparatus** (see Figure 15), does.

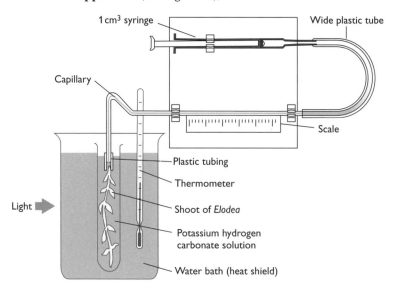

Figure 15 The Audus apparatus for measuring the amount of gas produced by a photosynthesising aquatic plant

Gas from the plant collects in the flared end of the capillary tube and, after a suitable time period (say, 10 minutes), is drawn into the capillary tube by withdrawing the plunger of the syringe. The length of the gas produced is measured against the scale. The volume can be determined if the diameter of the capillary bore is known (volume = length $\times \pi r^2$). To investigate the effect of light intensity, measurements are taken when a light source (e.g. a bench lamp) is placed at varying distances from the plant. The relationship between the distance of the lamp from the plant and the light intensity is described by the 'inverse square law'. The light intensity is inversely proportional to the square of the distance, $I \propto 1/d^2$ where I is the light intensity and d is the distance between the light source and the plant.

A graph of volume of gas produced in a standard time against the light intensity is used to present the results. The graph should show that at low light intensities, an increase in light intensity increases the amount of gas produced. It should also indicate **saturation**, that is at high light intensities, an increase in light intensity does not further increase gas production.

Referring to gas production as photosynthesis should be avoided for two reasons. Firstly, the gas does not necessarily consist only of O_2 (some nitrogen will be released by the surrounding water). Secondly, the O_2 produced can only ever be a measure of **net photosynthesis**, since some O_2 is used in respiration. Furthermore, there are a number of precautions and controls that need to be attended to:

- The apparatus has water run through it to expel any air bubbles before using it.
- Potassium (or sodium) hydrogen carbonate is added to the test tube to ensure that CO_2 is not limiting during the course of the experiment.
- The test tube containing the plant is immersed in a beaker of water, which acts to absorb any heat given off by the lamp.
- The room should be darkened during the experiment to exclude any other sources of light.
- After repositioning the light source, a period of 5 minutes is left before collecting gas to allow the plant to adjust to the new light intensity.

Question 13
Apart from producing O_2, an aquatic plant will take up CO_2. Explain what should happen to the pH of the surrounding water.

Demonstration of a plant's compensation point

At the **compensation point**, the rate of photosynthesis of a plant equals its rate of respiration. Its CO_2 input and output are equal. The **light compensation point** of a plant is the light intensity at which the rate of CO_2 uptake in photosynthesis equals the rate of CO_2 produced in respiration. Hydrogencarbonate (bicarbonate) indicator solution is sensitive to the amount of CO_2 in the air (since CO_2 is mildly acidic). At normal atmospheric levels the indicator is a red colour. If CO_2 is added the indicator turns yellow, while if it is removed from the air the indicator turns purple. This is illustrated in Figure 16. If the indicator remains red (in 'dim' conditions) then the plant is at its light compensation point.

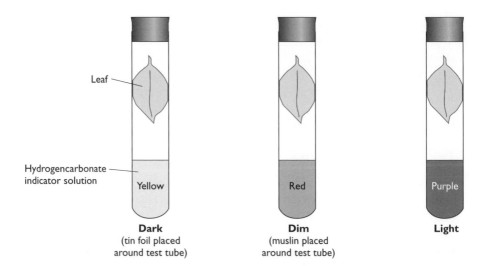

Figure 16 Using hydrogencarbonate indicator solution to demonstrate the compensation point of a leaf

Question 14

Look at Figure 16. Explain the colour of the indicator solution in 'dark' and in 'light' conditions.

Transpiration in plants (AS 2)

Measuring the rate of water uptake using a bubble potometer

A potometer (see Figure 17) is an apparatus for measuring the rate of water uptake by a leafy shoot. The water in the xylem vessels of the shoot is continuous with the water filling a capillary tube attached to its cut end. The rate at which a bubble of air moves along the capillary tube indicates the rate of water uptake. The major influence on the rate of water uptake is the rate at which water evaporates from the shoot in the process of transpiration. The usual way to take measurements is to record the scale reading initially and then subtract it from the reading after a set period of time (say, 1 minute) — this gives the rate of water uptake (mm min^{-1}). If the bore diameter is known, this can be expressed as a volume per unit time (mm^3 min^{-1}).

The potometer is used to investigate factors influencing transpiration, for example air currents, temperature, humidity and light intensity (influencing stomatal aperture). The influence of air currents can be investigated using a fan; for a control, no fan would be used. To investigate the influence of light, the plant could be enveloped by a black plastic bag (stomata should close). However, the covering bag would also result in an increase in humidity; the control experiment should involve the use of a transparent plastic bag.

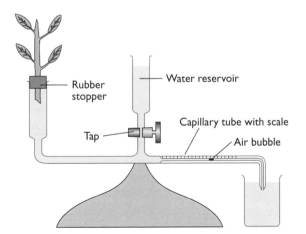

Figure 17 A potometer set up with a leafy shoot

It is imperative that an 'air lock' does not occur at the cut end of the shoot. The shoot is cut under water, the apparatus is assembled under water, the shoot is chosen to fit snugly into the rubber stopper of the potometer and the joint is sealed with grease. To compare the results for different plants, for example lilac (possessing leaves with thin waxy cuticles) and laurel (leaves with thick waxy cuticles), the results could only be compared if the surface area of the leaves was determined. Results would then be expressed in the form of volume of water taken up per unit leaf surface area per unit time ($mm^3 cm^{-2} min^{-1}$).

Question 15

Suggest how you might determine the surface area of the leaves of a shoot.

> **Tip** You are not expected to use the respirometer, the Audus apparatus or the potometer. The specification asks that you *understand* them, so a demonstration of their use is sufficient.

Ecological techniques (AS 2)

Ecological investigations are usually aimed at finding out something along the following lines:
- What types of organism are found here?
- How abundant are they?
- Where are they most abundant within the area?
- What abiotic (environmental) factors might be responsible for the distribution?

Sampling procedures

When you wish to investigate a population of organisms in a habitat it is not practical and too time-consuming to study every individual so you need to study a **representative sample** — a sample that is typical of the whole population. The type of sampling procedure will depend on what you are trying to achieve.

Random sampling

To study a plant or animal population in a reasonably uniform habitat, such as a meadow, you need to use a sampling device. For plants and sessile (non-moving) or slowly moving animals (such as limpets) this is often a quadrat (see below). This device is then positioned randomly throughout the area. In a random sample every individual has an equal chance of being included in the sample, so there is no bias. This avoids sampling only in the middle of the area, or selecting samples where you think the plant, or animal, might live. You should appreciate that a random sample does not guarantee a representative sample. It is possible, though unlikely, that a random sample of limpets on a rocky shore included all the tallest individuals. However, random sampling will provide unbiased information and is the best way in which to achieve a sample that is representative of the population in the area.

How many samples should be taken? Obviously, the more samples you take then the more reliable will be your estimate for the population. Generally in ecology, between 20 and 60 random samples are taken. This amount of replication is necessary because of the highly variable nature of any habitat. The actual amount of replication will depend on the amount of information assessed at each sample site: estimating the abundance of all species present is time-consuming and so 20 samples will suffice; determining the presence of a single species should not be too time-consuming, making possible a much higher level of replication.

How is a random sample achieved? Figure 18 shows how to place quadrat frames at random:

- Consider the area to be sampled as a grid with the bottom and left sides as axes — tape measures should be placed along these sides.
- Use a random number generator on a calculator to produce a pair of coordinates, for example X5, Y4 (this pair would define the bottom left-hand corner of the square on the diagram).
- Place a quadrat with its bottom left-hand corner on the intersection of the coordinates.
- Estimate the abundance of the species inside the quadrat.
- Repeat the random placement of quadrats an appropriate number of times.

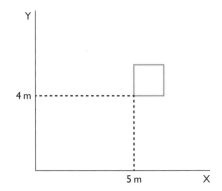

Figure 18 Placing a quadrat at random

Transect sampling

Transect sampling is used in the situation where there is a transition in the environment and zones of different species are present. These zones of different species types represent a **zonation**. Examples of **environmental gradients** may be observed in the following situations:

- the distribution of seaweeds and animals on a rocky shore
- the distribution of plant species at the edge of a lake
- the distribution of plant species from a meadow into woodland

Line transect

A **line transect** is set up across the areas where there are clear environmental gradients. This transect line is marked at intervals (say, every 0.5 m or 1 m) along its length. The species touching the line at marked intervals are recorded. The species can be presented diagrammatically as being either **present or absent** along the length of the transect. While information about species presence along the transect can be quickly obtained, the data produced by a line transect is somewhat limited.

Belt transect

A **belt transect** will supply more data than a line transect. It will give data on the abundance of individual species at different points along the line of transect. In a belt transect, the transect line is laid out across the area to be surveyed and a quadrat is placed on the first marked point on the line. The plants and/or animals inside the quadrat are then identified and their abundance estimated. Animals (such as limpets on a rocky shore) can be counted, while it is usual to estimate the percentage cover of plant species (see below).

Quadrats are sampled all the way along the transect line. If the line of transect is relatively short, say 20 m or less, then 1 m × 1 m quadrats might be placed **contiguously** (end-to-end) along the transect. If the length of the transect is much longer, then an **interrupted belt transect** is used: the quadrat is positioned at intervals, such as every 2, 5 or 10 m depending on the overall length to be sampled.

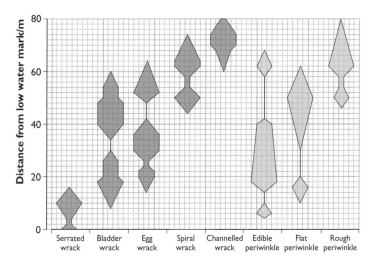

Figure 19 A kite diagram showing the distribution of seaweeds and some animals on a rocky shore

Data from a belt transect can be used to construct a kite diagram. The kite diagram in Figure 19 shows the distribution of seaweeds and some animals on a rocky shore. For each species, abundance is shown symmetrically round a value of zero. The data were obtained by placing a quadrat at the low water mark and sampling every other metre up the shore.

Question 16
A meadow slopes down to a wetland area. Explain the most appropriate sampling procedure for investigating the gradation in the vegetation down the slope.

Sampling devices

Quadrats
A **quadrat** is an area in which *plants or non-mobile animals* can be sampled. Quadrats are usually square: ½m × ½m and 1 m × 1 m are common but larger quadrats can be constructed. Generally, smaller quadrats are used to sample smaller plants and those that are common in the area sampled; for larger plants and for less common plants a larger quadrat size is required. Also, less common species require a greater number of samples (since the species might be expected to be absent from many of the quadrats).

To *estimate the abundance of a species* (plant or non-mobile animal) in an area:
- determine the abundance in a number of quadrats (say 30 ½m × ½m quadrats)
- calculate the mean abundance per quadrat (for a ½m × ½m quadrat this would be per 0.25 m²)
- estimate the total area within which sampling has taken place
- determine the overall abundance by dividing the mean abundance in a quadrat by the quadrat area and multiplying this by the total area

Pitfall traps
A **pitfall trap** is used to capture arthropods (insects, spiders, etc.) *walking over the soil surface*. A jar is inserted in the soil so that its rim is flush with the surface, and roofed to prevent it flooding with rainwater (see Figure 20). Passing arthropods, such as beetles, fall into the jar and cannot escape. The jar should be examined at frequent intervals (to reduce the likelihood that carnivorous arthropods eat the others) and living organisms removed for identification and counting.

Pitfall traps may be randomly arranged in a habitat to determine the types of arthropod there or to estimate the relative density of a particular species of beetle; or they may be arranged along a transect through habitat boundaries, such as through the edge of woodland leading into a meadow.

Roof to keep rain out

Supporting pebble

Container sunk into soil

Figure 20 A pitfall trap, used to sample arthropods moving over the soil surface

Sweep nets and pooters

A **sweep net** is a net used for sampling arthropods *from tall grassland*. A standard number of sweeps, say 10 sweeps from one side to the other, is undertaken. The arthropods collected are trapped in a **pooter** (see Figure 21). The organisms trapped in the sample are identified and counted. This is a particularly useful technique for determining, say, the relative abundance of an insect, inhabiting a meadow, from spring through to autumn. (The data in Figure 24 on page 50 were obtained using a sweep net and pooter.)

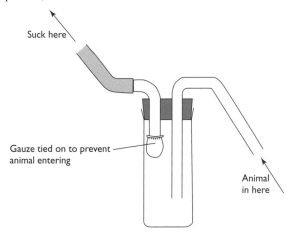

Figure 21 A pooter, used to suck up small arthropods collected in a sweep net

Question 17

A lawn, 9 metres wide by 20 metres long, was sampled using 30 ½ m × ½ m quadrats to estimate the percentage cover of creeping buttercup (*Ranunculus repens*). The average percentage cover per quadrat was found to be 2%. Determine (a) the percentage cover of the whole lawn, and (b) the total area of the lawn covered by creeping buttercup.

Estimation of species abundance

Abundance relates to the amount of a species in an area. There are different measures of abundance such as density, percentage cover and frequency.

Density

Density is simply the *number of organisms per unit area*. This is appropriate for determining the abundance of animals, such as limpets (on a rocky shore), which move very slowly and are not hidden, or for plant species in which individuals are easily recognised, such as orchids. The number of individuals of a species within a quadrat is counted and, after repeating this in different positions, the results can be expressed as the average number of individuals per unit area — the density.

Percentage cover

Often it is difficult, with plants, to count individuals. For example, do the blades of grass belong to one individual or several? Also, plants often differ greatly in size and so counting numbers is not necessarily a good estimate of abundance. For example, on the rocky shore there may be small and very large specimens of the seaweed egg wrack (*Ascophyllum nodosum*). To get around this, you should estimate the **percentage cover** of a species.

Using a quadrat, you estimate the portion of the area covered by the plant or seaweed species under investigation. If it covers a quarter of the quadrat then the percentage cover is 25%. A quadrat subdivided by a grid into smaller squares will help in estimating % cover; in a 5 by 5 grid, each square is 4% of the total area (see Figure 22). If the species covers 6 whole squares then the percentage cover is 6 × 4% = 24%. If the species covers 4 whole squares and parts of other squares that are equivalent to 3 squares then a total of 7 squares is estimated, giving 7 × 4% = 28%.

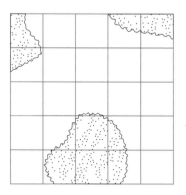

Position of creeping buttercup

Figure 22 The cover of creeping buttercup (Ranunculus repens) in a quadrat

Question 18
Figure 22 shows the cover of creeping buttercup in a quadrat. Determine the percentage cover of the creeping buttercup in the quadrat.

Frequency
The determination of frequency uses **presence-absence data**. Is the plant present or is it absent? Data may be collected by sampling along a line transect or by looking for presence in a series of quadrats (or point frames, another sampling device). Frequency is calculated as *the number of times that a particular plant species is present at sampling points divided by the total number of sampling points* — described sometimes as the total number of 'hits' divided by the sum of the number of 'hits' and 'misses'. Frequency is not a measure of the abundance at any sampling point and so provides rather limited data. Its use is in areas where there is a lot of data to obtain and determining presence or absence allows this to be completed quickly.

Influence of environmental factors on the distribution of organisms

At each sample point, the environmental factors that might influence the distribution of the organisms might be monitored. **Abiotic factors** might include:
- light intensity
- soil water content
- soil pH
- nutrient availability in the soil

} factors relating to the soil are called **edaphic factors**

Tip You are not expected to *know how* to measure any particular environmental factor, but you are expected to *appreciate* that they can be measured. You can always look up how to measure any abiotic factor in an investigation of plant distribution. For example, nettles (*Urtica dioica*) appear to grow most densely in soils with high phosphate levels; while the meadow buttercup (*Ranunculus acris*) tends to be more frequently found in damp meadows — these might make useful A2 investigations.

You can expect to undertake **fieldwork** to collect data on populations of plants and/ or animals in their environment. You are expected to know about sampling, sampling devices and measures of abundance (as presented in this guide). You are also expected to be able to explain your results or, at least, make reasoned suggestions, and evaluate the procedures used and the data collected. When evaluating ecological tasks expect to comment on the following problems:

- sampling may be biased, so that the results are not representative
- too few samples may be taken, so that the results are not reliable
- estimating the abundance of different species may be difficult, so that errors decrease the accuracy of the results
- identifying different species may be difficult, so that misidentification reduces the validity of the conclusions

Demonstrating practical skills at AS

At AS, you will be assessed on *Practical Tasks*, which should include the following skill areas:

- **Implementing** a sequence of instructions
- **Recording and communicating** results
  - **Tabulation** of raw and derived data
  - **Graphical** presentation of the data
- **Interpretation** of the results
- **Evaluation** of the experimental design

Implementation

Implementing practical work requires that you have had practice with a range of techniques, and that you have developed the ability to work safely and understand the instructions to be followed so that you can work in an organised fashion. You must be able to take precise and accurate measurements, and understand that repeating results allows you to estimate reliability.

Developing skills in the use of apparatus and materials

You should have carried out appropriate practical techniques before undertaking any of the assessed practical tasks. There is a range of practicals designated in the specification (and described in the previous section of this guide). Use these experiences as opportunities to develop your skills in handling apparatus and associated materials. You should be able to carry out a wide range of manipulative techniques with a high degree of skill. These include the skilful use of laboratory equipment and the accurate measurement of, for example, volumes, masses or temperatures during the experiment.

Working safely

Here are some simple rules to follow to make sure that you work safely in the laboratory:

- No eating or drinking in the laboratory.
- Always move slowly in a laboratory.
- Wear a protective lab coat and goggles where necessary.
- Familiarise yourself with the location of first aid equipment.
- Tie long hair back, particularly if there are any naked flames.
- Tuck your bags and other personal items safely under the bench before you begin.
- Consider all chemicals as potentially dangerous (corrosive, toxic) and microbial cultures as potential health hazards, unless you are told otherwise.
- Use aseptic techniques with microbial cultures.
- Wash your hands on completion of practical work.
- Never place your fingers in your mouth or near your eyes after using chemicals or touching biological specimens.
- Never pipette by mouth — use a pipette filler.
- Make sure that fragile objects such as glassware are placed where they cannot be knocked over or rolled off the bench.
- Allow hot objects such as Bunsen burners, tripods, gauzes and beakers to cool down before you handle them.
- If you are not sure how to do something, ask for help.
- Report all accidents and potential hazards to your teacher, for example faulty equipment, breakages, spillages.

Safety and care with living organisms

Biology practicals may involve working with living organisms. This will undoubtedly be the case when undertaking ecological work outside the laboratory, i.e. in the 'field'. Safe work with, and care for, living organisms requires that you:

- take measurements of plants *in situ* and do not damage or remove them
- handle living animals in a way that avoids any distress
- wash your hands with soap and water after completing fieldwork

Risk assessments

The instructions for your practicals should include risk assessments. Make sure that you read the risk assessment before starting and that you take adequate precautions to minimise any risks. The risk assessment will tell you if you are going to be using any chemicals that are hazardous. It may be necessary to use substances that are corrosive (such as potassium hydroxide), highly flammable (such as ethanol) or harmful (such as ethanal). The risk assessment should include suitable control measures (safety precautions) and recommend a procedure for disposing of the solutions at the end of the practical.

Understanding the procedures and organising work

Ensure that you have a full understanding of the procedure before starting any practical work. One of the commonest ways for a practical to go wrong is starting one step of the procedure only to have to pause because some other step should have been prepared beforehand. For example, water baths may need to be set up early so that they are at the correct temperature when you go to use them; some colorimeters need to be switched on beforehand so that they 'warm up'; while a chromatography tank containing solvent will need to be left for some time while the atmosphere within becomes saturated with solvent vapour.

So read the procedure carefully. Rewriting the procedure as a flow diagram can be a helpful device in improving your understanding.

Using your list of apparatus and materials, check that everything you need is gathered together on the bench. Push the items that you will not need at the start towards the back of the bench. As part of your preparation label all the tubes and containers. Think about how you are going to record your data and then prepare some sheets, so that you are ready to write down your measurements and observations.

Make sure that you give yourself time to carry out the investigation. Students often become careless when they are rushing to complete something.

Understanding variables

While reading through the procedure, identify the variables. In laboratory practicals there will be an independent variable, a dependent variable and other variables that need to be controlled. Sometimes, in ecological investigations in the 'field', an independent variable is not obvious, while it is difficult to control other variables.

Independent and dependent variables

The **independent variable** (IV) is that factor which is being manipulated or changed by you during the practical; the factor that is consequently measured is the **dependent variable** (DV). You might ask yourself which variable is causing a change in the other? In this case the IV is causing a change in the DV. The experiment is trying to establish cause and effect. For example, in an experiment to find the effect of temperature on the rate of enzyme activity, the temperature would be varied, say, from 10°C to 60°C in steps of 10°C. For each temperature the rate of enzyme activity would be measured using a suitable technique. The temperature is thus the IV since it is the one that is deliberately being varied by the experimenter, and the rate of enzyme activity is the DV. If you are unsure about which is the dependent and which is the independent variable try completing this sentence:

'The [insert the DV] depends on the [insert the IV]'

Interdependent variables

On occasions, data will be obtained in an attempt to find out whether there is a relationship between two variables. Where changes in a variable are connected with changes in another variable, but there is no hint of independence or dependence, the relationship is **interdependent**. For example, this is the case with the size of heather (*Calluna vulgaris*) plants and the pH of the soil in which they were growing. It is not possible, without further experimentation, to determine whether it is the heather plants that are influencing the pH of the soil or if the soil pH influences the growth of the heather (or a combination of both).

An understanding of the relationship between these variables is important when you come to record your results in a table and when you decide which is the most appropriate form of graph to use in presenting your results.

It is also important to identify those variables that might influence the results and so need to be kept constant, i.e. controlled. These are the **controlled variables** and they allow for a *fair test*. How well have these variables been controlled? Are there some that have not been adequately controlled? You should ask these questions at an early stage because you will need to discuss such issues during your *interpretation* of the results and/or your *evaluation* of the experiment.

Making measurements: precision, accuracy and errors

You will be making measurements during the course of your practical. This will include your measurements of the DV. You will be expected to be precise and accurate in measuring the following:

- **Mass**: make sure that the pan of the electronic balance is clean and that the reading on the balance is at zero before you take any measurements.
- **Volume**: make yourself familiar with measuring cylinders, burettes, graduated pipettes and syringes to measure out precise volumes of liquids; when taking readings of levels of liquids you must line up your eye with the level of the liquid, and read the volume against the bottom edge of the meniscus.
- **Length**: you will need to take readings from a ruler scale alongside a capillary tube when using a respirometer, potometer or Audus apparatus, and you will need to measure the length of the air column in a J-tube. It is easy to introduce a source of error if you vary the position of your head relative to the scale when you make a measurement, so you must always view the scale so that your line of sight is perpendicular to it.
- **Time**: if you are measuring how much something happened in a 10-minute period then you should make sure each measure is precisely 10 minutes; or you may need to use a stopwatch to measure the time it took for some event to be completed, in which case you need to be familiar with the working of the stopwatch in use.
- **Temperature**: you may need to check the temperature of a solution, in which case always stir the fluid before taking a thermometer reading.

The measurements that you make can only ever be estimates. Obviously, you aim to ensure that your estimate is close to the 'correct' value (i.e. the theoretical true value). However, there are a number of reasons for your results to deviate from the correct values and you must consider these, both during your *implementation* and, later, when you write an *evaluation* of your experiment.

Precision

There may be limitations in the measuring instrument. For example, the scale on a colorimeter (though many are now digital) or along the calibrated part of a potometer will have divisions of finite width. This means that readings are taken to the nearest unit. The result is a lack of precision. So **precision** is related to the *smallest scale division on the measuring instrument* that is being used.

Accuracy

The *instrument that you use to make your measurement* will influence the **accuracy** of the measurement. The level of accuracy required depends on the magnitude of what is being measured. For example, if you are timing a reaction that is likely to last only a few minutes, a stop clock is the best choice. If you are timing something that lasts several hours, that level of accuracy is not needed. As a 'rule of thumb', measuring to 0.1% accuracy is enough for your purposes. As a further example, you have a choice of instruments for measuring volume: a selection of measuring cylinders, pipettes and syringes. If you are required to take a sample of volume 1 cm^3 then using a 1 cm^3 syringe will provide a more accurate measure than a 10 cm^3 syringe or 10 cm^3 measuring cylinder. However, to measure 200 cm^3 you would be better to use a 250 cm^3 measuring cylinder (reading the volume at the 200 cm^3 mark, of course).

Errors

Errors during an experimental procedure can be systematic or random. An error is **systematic** if the results are regularly too big or too small — there is said to be **bias** in the data. For example, in determining the water potential of potato tissue, discs of tissue are surface dried before weighing, initially (before immersion in a sucrose solution) and finally. If the final masses are measured with discs that have been insufficiently dried, then all measures of 'changes in mass' will be overestimates — a systematic error. An error is **random** if the measurements made are sometimes too large or sometimes too small, without being able to predict which is happening. For example, when using the iodine test to determine when starch has been digested you have to decide when a blue-black colour has no longer been produced. This involves personal judgement (that is, dependent on the subjectivity of the student) and will introduce a degree of variability in your data.

The problem of errors can be kept to a minimum by careful selection of material and the careful control of variables. Procedures should be standardised. In the experiment for determining the water potential of potato tissue you would have standardised your method of surface drying each disc of tissue before measuring its mass.

Repeating measurements: replication and reliability

Biological processes are most often influenced by a range of factors. Of course, you will have identified those variables that you need to control (the controlled variables). Keeping these factors constant is not always easy. For example, in an experiment on the influence of temperature on enzyme activity you will use a range of water baths. However, while any one water bath will be set (and checked) as being at a certain temperature (say, 30°C), the temperature will in fact fluctuate around a particular value. Also, in an experiment on the influence of light intensity (determined as the inverse of the distance of the lamp from the plant) on the production of gas by an aquatic plant, factors may change during the course of the experiment: the temperature of the water, the amount of available carbon dioxide in the vicinity of the plant and the amount of extraneous light. Furthermore, biological material is notably variable. For example, the water potential of potato tissue can be determined by immersing cylinders of tissue in a range of concentrations of sucrose solutions. However, different pieces of tissue will vary in their water potential especially if they have been taken from different potato tubers. The result is that in biological investigations, differences in the material used or changes in the conditions in which they are carried out, can cause a lot of variation in results.

Ultimately, you must expect that repeated results will vary. These repeated results (which may be produced by different members of the class) are called **replicates**. Replication allows you to estimate the **reliability** of the data. Reliability is a measure of the closeness of agreement between individual results. The lowest level of replication is three. Obviously, increasing the number of replicates (i.e. the sample size) improves your ability to estimate reliability. Sample size — the number of replicates to be used — is discussed further when you undertake a planning exercise at A2 (see page 84). There are statistical methods, which you will use at A2, to describe the variation in replicate results, that is, the reliability (see page 80).

Repeating the experiment many times also allows you to:
- spot odd results (**anomalous results**) and, if justified, to exclude these
- calculate an arithmetic average (**mean**), which is likely to be more representative than any individual result

While experimental results may only be estimates, it is important that you:
- make the best effort to avoid errors in the design of investigations and the use of instruments
- are aware of the sources of errors and appreciate their magnitude

All your readings — the raw data — should be noted on your record sheets. In addition, you need to make notes of anything unusual that occurred and/or any problems that you encountered.

Recording and communication of results

Biology experiments generate a lot of data. It is important, first of all, to understand something about the nature of the data. Data can be:

- **quantitative (numerical)** — the data are measured or counted on a numerical scale. Quantitative data can be:
  - **continuous** — the data can take on any value in a certain range, e.g. temperature values
  - **discontinuous** or **discrete** — the data take on only certain numerical values, e.g. number of offspring in different families (0, 1, 2 , 3, etc.)
- **Qualitative (categoric)** — the data describe attributes or categories, e.g. colour as red, yellow, green or blue.

Quantitative data may have been obtained as a result of *measuring* (e.g. length, volume, time, transmission) or *counting* something. So they are either:

- **measured data**, e.g. the volume of oxygen produced by pondweed (*Elodea*) in a period of time; or
- **frequency data** where counts have been made, e.g. number of plasmolysed cells out of a total number viewed

Constructing tables to record results

The most convenient way of recording data is in the form of a table.

Raw and processed data

The **raw data** are those that are immediately recorded. In some cases they need to be processed. Three examples of **processed data** (also called **derived data**) are given below.

- Data in which the **initial value is zeroed** and subtracted from other values. For example, an experiment using a potometer (to compare transpiration in 'normal' room conditions with transpiration in air currents produced by a fan) might generate figures (on the left side) which would need to be processed (on right side) to facilitate comparison:

  'normal' conditions   13, 17, 21, 25, 29   →   0, 4, 8, 12, 16
  in air currents (fan)   8, 15, 22, 29, 36   →   0, 7, 14, 21, 28

  It is now easier to compare the two sets of data. Subtracting the initial value may also be required when taking measurements using a respirometer or an Audus apparatus.
- Data in which the change in mass (or length) is better shown as a **percentage change** in mass (or length). The reason for this is again to allow direct comparison of the data, which would not be the case if the initial masses were different (even slightly different). This is most obviously the case when using data to determine the water potential of a 'bulky' plant tissue (such as potato tuber).
- Data that are measured on a time scale and then **converted to a rate**. For example, in an enzyme experiment you may have measured the time taken for

digestion to be completed. This can be converted to a rate by calculating its inverse as follows:

$$\text{rate of reaction} = \frac{1}{\text{time taken for digestion to be completed}}$$

It is better to show the rate of reaction (if this is possible) since it is directly related to the activity of the enzyme and so makes interpretation easier.

In *assessed experimental tasks* (or in the A2 investigation), you will have obtained a number of replicates or repeats (at least three) of the measurements. The replicated data are used to **calculate a mean** and to assess the variability of the results and so aid in making a decision about the reliability of the results. Both the replicated data (the repeats) and their means should be shown in any table. It will also be possible to identify any anomalous results.

Alternatively, if the data represent counts (frequency data) then the results are summarised as **percentages of the total counts**. For example, the number of seeds germinated divided by the total number of seeds used, and then multiplied by 100, will give you a figure for the percentage germination. Also, in the experiment in which you have to determine the average solute potential of plant tissue you would count the number of plasmolysed cells and non-plasmolysed cells and convert the data to percentage plasmolysis.

It is important that careful consideration is given to the arrangement of the data in the table. Some time should be taken in deciding how you are going to organise the data. This should be done prior to undertaking the practical rather than after results have been produced. The arrangement needs to make it easier to:

- identify anomalous results and make judgements about their reliability
- plot an appropriate graph
- identify any trends

Some of the important steps in the construction of a table are shown in Table 3.

Table 3 Important steps in the construction of a table.

Tabular feature	Description of process
The table should have a caption (also called a legend or a title).	This must include the independent variable (e.g. 'the effect of X on...'), the dependent variable (the process being affected) and the biological material being investigated.
There should be a suitable number of columns, each with an appropriate heading.	The first column should be that for the independent variable (i.e. the variable that you changed in the experiment). Subsequent columns should be those for replicates (and mean values) for the dependent variable (i.e. those that you measured).
The column headings should include the units of measurement (as appropriate).	The heading is followed by a solidus (or slash) and then the unit of measurement. SI units should be used — see Table 4.

Tabular feature	Description of process
There should be an appropriate number of rows.	This should include the rows for headings as well as rows to cover the range of the independent variable and any totals.
The data should be added in a uniform way.	They should be consistent in the number of significant figures (often three significant figures, e.g. 6.34).
	Very large or very small numbers are more easily expressed in standard form (e.g. $3\,750\,000$ as 3.75×10^6 or 0.000271 as 2.71×10^{-4}).

Units of measurement

It is important that you know the units of measurement used in practical work. Table 4 shows the units that you are most likely to use.

Table 4 Common SI units.

Unit	Name of unit	Symbol	Notes
Time	Second	s	While second is the SI unit, others are acceptable. In most laboratory experiments you are likely to measure in minutes.
	Minutes	min	
	Hour	h	
Length	Metre	m	The centimetre is not an SI unit, so most measurements are given in millimetres.
	Millimetre	mm (10^{-3} m)	
	Micrometre	μm (10^{-6} m)	
	Nanometre	nm (10^{-9} m)	
Volume	Decimetre cubed	dm^3	These are the units used in examination papers, although you may well find litre (l or L) or millilitre (ml) scales on apparatus such as glassware and syringes.
	Centimetre cubed	cm^3	
Amount of substance	Mole	mol	Concentrations may be given as mol dm^{-3} (often said as moles per litre) or as mass per volume, e.g. g dm^{-3} (grams per dm^3); you will also see percentage solutions, which are g 100cm^{-3} (grams per 100 cm^3)
	Millimole	mmol (10^{-3} mol)	
	Micromole	μmol (10^{-6} mol)	
Temperature	Degrees Celsius	°C	–
Pressure	Pascal	Pa	Pressure units are used for water potential, solute potential and pressure potential in experiments on osmosis.
	Kilopascal	kPa (10^3 Pa)	
	Megapascal	MPa (10^3 kPa)	

Format of the table

Table 5 shows a table prepared to record the results from an investigation in which the effects of two types of amylase are studied over a range of temperatures.

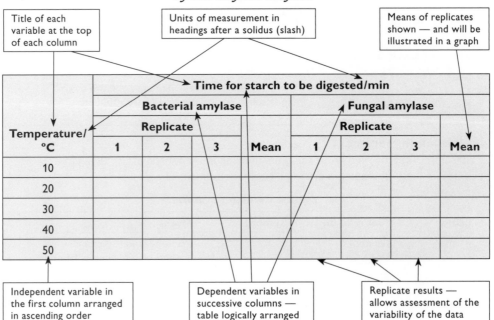

Table 5 The effect of temperature on the digestion of starch by two amylase enzymes

Title of each variable at the top of each column

Units of measurement in headings after a solidus (slash)

Means of replicates shown — and will be illustrated in a graph

Temperature/ °C	Time for starch to be digested/min							
	Bacterial amylase				Fungal amylase			
	Replicate				Replicate			
	1	2	3	Mean	1	2	3	Mean
10								
20								
30								
40								
50								

Independent variable in the first column arranged in ascending order

Dependent variables in successive columns — table logically arranged

Replicate results — allows assessment of the variability of the data

Anomalous results

An anomalous result is one that seems to be out of line with the others. You may have repeated the experiment four times and have three replicates that generated similar measurements, but one replicate with very different values. What do you do if an anomalous result is detected? This depends on the situation.

- If during the implementation of the experiment you notice that one (or more) of the measurements looks out of line, then you have the opportunity of repeating that bit of the experiment.
- If an anomalous result does not become apparent until you analyse the data (and implementation is complete) then you should identify the anomaly and, if you can suggest a reason for it, you may ignore it in subsequent analysis. This is not to ignore the anomaly but to ensure that it does not distort the overall trend. If you cannot suggest an explanation for the anomaly then you will just have to put up with it.

Assessing the variability of the replicates: assessing reliability

You will be reporting on the variation shown among the replicates and so assessing the reliability of the results in the 'evaluation' section of your report, but it is important to make initial observations at this stage.

- If the values among the replicates are all rather close so that there is **little variation** then the results may be regarded as reliable; in this situation, you can be more confident about drawing conclusions from the results.
- If the values among the replicates are rather spread so that there is **greater variation** then the results cannot be regarded as being reliable; in this situation,

you should attempt to suggest reasons for the greater spread and be a little tentative about any conclusions made.

When assessing the variability of the results, it is important that you do not just refer to the highest and lowest values (the **range**) since you are only looking at two values. You must consider all the values. (There will be an opportunity to assess reliability using statistical analysis in the A2 coursework.)

Graphical presentation of the results

Once your table is complete, the set of data can then be presented as a graph. A graph is an illustration of how variables relate one to another. It is basically a 'picture' of the results. It allows the trend or pattern in the data to be more easily seen. Graphs are an aid to understanding and interpretation.

Different types of graph are available depending on the nature of the variables and on the nature of the data. There are *four main types of graph* that you might construct in AS biology:

- **line graphs**
- **histograms**
- **bar graphs (bar charts)**
- **scatter graphs (scattergrams)**

The appropriate use of the different graphical techniques is summarised in Table 6.

Table 6 The appropriate use of different graphical techniques

Graphical technique	Independent variable (IV)	Dependent variable (DV)	Appropriate use	Example
Line graph	Continuous data	Continuous data	To determine the nature of a causal link between the IV and the DV	The effect of pH on the rate of reaction of two protease enzymes
Histogram	Continuous data that have been subdivided into classes	Quantitative data (which may be discrete — frequencies)	To illustrate the frequency distribution of the IV in a population	A frequency distribution of leaf width of wild garlic (*Allium ursinum*) leaves
Bar graph (bar chart)	Qualitative data — different attributes or categories	Continuous data	To illustrate the effect of different attributes on the DV	The effect of different colours of light on the rate of photosynthesis
Scatter graph (scattergram)	Both variables have continuous data though there may not be an obvious independent variable, i.e. they may be interdependent		To determine the relationship between two variables where there is not necessarily a causal link	The relationship between soil pH and the diameter of heather (*Calluna vulgaris*) stems

Irrespective of the type of graph, there are guidelines concerning its construction; these are summarised in Table 7.

Table 7 General guidelines for the construction of graphs

Feature	Description
Caption or title	This must include the independent variable (e.g. the 'effect of X on...'), the dependent variable (the process being affected) and the biological material being investigated.
Axes right way round	The independent variable (if known) must be presented on the x-axis.
Appropriate scaling	Appropriate scales for both axes should be devised to make maximum use of the graph paper — the graph should fit on the paper yet be sufficiently large to allow a more accurate plotting of results and facilitate interpretation. A scale of three measured units to one unit on the graph paper should be avoided as this makes plotting more difficult. Furthermore, the data should be critically examined to establish whether it is necessary to start the scale at zero.
Axes with labels and units of measurement	Each axis should be clearly labelled. Each label should be followed by a solidus (or slash) and then the unit of measurement (if appropriate). The axes may be offset to prevent data points being obscured by the scale line (see Figure 24, page 50).
Accurate plotting	All points or bars (columns) should be plotted accurately. It is best that a sharp pencil is used so that errors can be readily corrected. If two sets of data are plotted on the same axes then each set needs to be labelled or a key added.

The caption, along with axes labels, are important since someone should be able to look at the graph and know exactly what it is showing without any further explanation.

Line graphs

Joining the points

Points may be notated with a saltire cross (✗) or an encircled dot (☉ — though it is possible to have a triangle or square around the dot, or have the symbol filled in). Both can be used when two or more sets of data are to be plotted on the same set of axes. It is only for these points that measurements have been taken. It is not actually known what is happening between the points, so why join them up? Basically, there are two reasons for joining up the points:

- It makes it easier to see the trend or relationship between the two variables.
- It may be important to determine what is happening between points — a process known as interpolation — in order to make a calculation.

Once the points are plotted a decision must be made as to whether to draw a line of best fit (or trend line) — either a straight line or, if theoretical considerations allow or the general trend of the points indicates, a smooth curve — or a series of short straight lines joining the points. Conventions for A-level biology have been established by the Society of Biology, which provides the following guidance:

- *A smooth curve can be drawn to represent the relationship when there are sufficient data points to be confident in the relationship or, because from theory, there is good*

reason to believe that the intermediate values fall on the curve, e.g. the course of an enzyme-controlled reaction.
- When there are insufficient data to be able to confidently interpolate the relationship, straight lines joining the points should be drawn, thus indicating uncertainty about the intermediate values, e.g. numbers of ground beetles found at set distances from a hedge.

In most cases, it is better to join the points with straight lines because you do not know how the values between the recorded points may vary. Use a ruler to produce the straight lines.

Figure 23 shows line graphs of data for the production of gas by *Elodea* (an aquatic plant) at different relative light intensities (calculated as the inverse of distance of lamp from the plant). Short straight lines are the preferred method of joining the points and, while a smooth curve might be acceptable, it is less easy to achieve — see Figure 23(a). What is not acceptable is an undulating curve or a smooth curve that is not best-fit — see Figure 23 (b). Note also that it is not appropriate to continue the line to the 0, 0 coordinates ('origin').

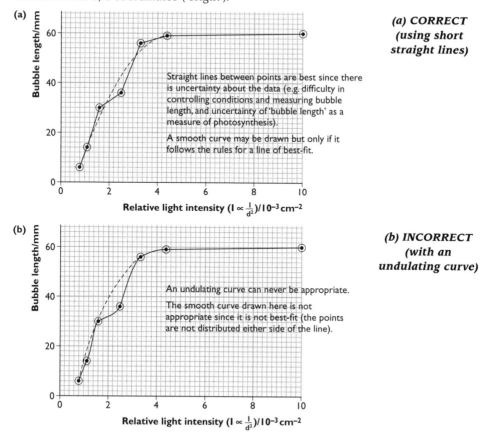

Figure 23 The effect of light intensity on the production of gas by Elodea

Figure 24 shows the results of sampling two species of insect with a sweep net during one summer in a meadow at Kinnego Bay. Notice that drawing smooth curves would be totally inappropriate. The data are prone to experimental error and so there is uncertainty about the results.

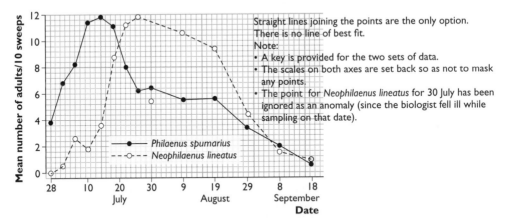

Figure 24 The relative frequencies of the adults of two spittlebug species (Philaenus spumarius and Neophilaenus lineatus) sampled with a sweep net during one summer in a meadow

Anomalous results

After plotting the points you can decide if any are anomalous. Ask yourself: 'do they fit the trend?' What should you do if you spot an anomaly? Unless you can account for the anomalous result (possibly due to some fault in the procedure or equipment) you are stuck with it. If you can explain the anomaly then the datum point may be ignored. This is the case shown in Figure 24, where the biologist fell ill and developed a high temperature on the day of sampling.

Lines of best fit

There is one situation where you must draw a straight line or a smooth curve (a trend line). It may be important to interpolate the data points in order to make a calculation. If interpolation between two points is required, accuracy is improved by drawing a line of best-fit so that all points are considered and not just the two points between which a calculation is to be made. There are a number of situations in which you have to make calculations from a graph. These include:

- constructing a calibration curve for converting colorimeter reading (% transmission) to percentage starch (using iodine to produce a blue-black solution); a *smooth curve* is required — see Figure 8, page 22.
- plotting percentage change in mass of potato tuber against water potential (or molarity) of the bathing sucrose solution and determining the water potential of the potato tuber where the line of best fit (a *straight line*) intersects the x-axis with no net change in mass — see Figure 10, page 23.
- plotting percentage plasmolysis against water potential (or molarity) of the bathing sucrose solution and determining the average solute potential of the

epidermal tissue from the line of best fit (most likely a *sigmoidal or S-shaped curve*) at 50% plasmolysis — see Figure 12, page 25.

A line of best fit is added *by eye*. To fit a straight line, you should use a transparent plastic ruler to aid you (and remember to use a pencil to allow correction). To fit a curved line, a flexicurve is useful, but if one is not available, a freehand curve should be drawn as smoothly as possible. When judging the position of the line the following rules should be applied:
- there should be approximately the same number of data points on each side of the line
- the points should be evenly distributed, either side of the line, both at the top and the bottom of the line
- the line should be near as many points as possible

It is important to emphasise that it is not necessary to connect any points when you are constructing a best-fit line. Certainly, do not simply join the first and last points.

> **Tip** For AS written papers, you may be required to plot a given set of data. The examiners have stated that students will not be penalised for joining points by short straight lines. Further, if a straight line or smooth curve is required for the purposes of interpolation (to make a calculation) then the question stem will direct you to draw a line of best fit. Remember, a line of best-fit must be drawn correctly, obeying the rules given above. So, *connect data points with straight lines unless told to do otherwise.*

Scaling

You do not need to start graphs at 0, 0. For example, if when measuring pulse rate before and after exercise all rates vary between 70 and 120 then you might well start the *y*-axis scale at 50 or 60, giving more space to plot the graph and making trends clearer — see Figure 25.

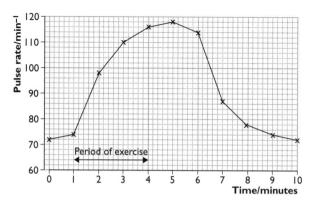

Figure 25 The effect of exercise on pulse rate

Sometimes two dependent variables are plotted on the same graph. If the scales for the dependent variables are quite distinct then two scales should be devised, preferably one on the left and one on the right — see Figure 26.

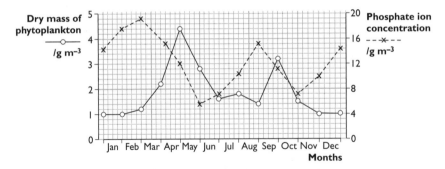

Figure 26 Annual changes in phosphate ion concentration and phytoplankton biomass in a lake

These aspects of scaling also apply to other graphical forms.

Histograms

A histogram is used to plot frequency distribution with continuous data. It would be used, for example, to show the frequency distribution of leaf width of wild garlic leaves. The leaf widths are grouped into classes. The frequency for each class is shown as a column, and adjacent columns should touch. An example is shown in Figure 27.

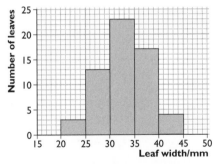

Figure 27 The frequency distribution of leaf width in a population of wild garlic (Allium ursinum)

Bar graphs (bar charts)

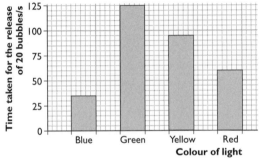

Figure 28 The effect of different colours of light on the time taken for the release of 20 bubbles by the Canadian pondweed (Elodea canadensis)

A bar graph is used when the independent variable is not numerical. The bars are of equal width and they do not touch. It would be used, for example, to show the effect of different colours of light on the rate of photosynthesis. In the experiment, shoots of the Canadian pondweed (*Elodea canadensis*) were exposed to light using various colour filters, and the times taken for the release of 20 bubbles from the cut end of the stem were recorded. The histogram is shown in Figure 28.

Scatter graphs (scattergrams)

A scatter graph or scattergram shows the relationship between two variables. The points on the graph are not joined up but left as simple crosses. It would be used, for example, to determine if there was a relationship between the size of the heather plants and the pH of the soil. In the investigation, 10 heather plants were chosen at random and their size estimated by measuring the diameter of the main stem, while the pH of the top centimetre of soil under the centre of each plant was also determined. The scattergram is shown in Figure 29. In this example, soil pH may be influencing the growth of the heather, but it might just as well be that the heather is causing a more acidic soil pH. The axes could, indeed, be drawn the other way round. There is no obvious IV.

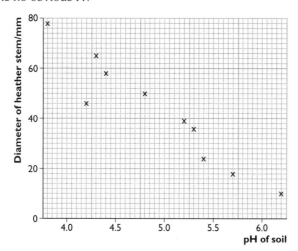

Figure 29 The relationship between soil pH and the diameter of heather (Calluna vulgaris) stems

The way in which the points fall on the graph shows the trend. Figure 30 shows three possible trends.

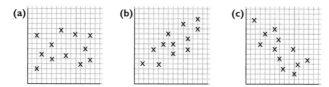

Figure 30 Three possible trends on a scattergram: (a) no correlation; (b) positive correlation; and (c) negative correlation

The stronger the correlation, the more the data points tend to be clustered along a line.

Other graphical forms

Two other graphical forms are to be found:

- **pie charts** — these show the data as portions of a whole
- **kite diagrams** — these show the relative density of a species along a transect — see Figure 19, page 33.

Interpretation

Identifying trends

Once you have produced your tables and graphs, take some time to study them carefully. Make sure that you understand the data concerned, read the captions carefully and make sure that you understand the labels on the axes. Concentrate on the overall shapes of the patterns or trends on your graphs. Do not worry about minor fluctuations. Some of the more common patterns are shown in Figure 31, along with the terms used to describe them.

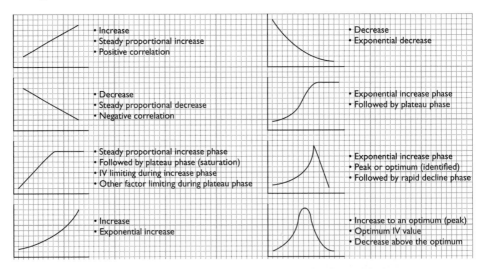

Figure 31 Some common trends in biology, with descriptive terms

Keep your description of the trends and patterns concise. In describing a trend, do not re-state every result — they should be recorded in the results table. However, it can be helpful to quote one or two figures to help the description. For example, at what point does the curve level off or what is the optimum value? If you have several curves plotted on the same graph, try to pick out comparative points. What have the curves got in common and how do they differ? Remember that you are aiming to produce an overview or a summary.

Tip Be careful when describing trends that you don't describe changes in relation to 'time' when time is not the independent variable. For example, the last graph shown in Figure 31 is the type of trend that you might expect from the influence of pH on enzyme activity. It is not acceptable to say that the curve rises initially, peaks and finally falls. Rather, the trend shows that from a low pH (identified), as pH is increased activity increases to an optimum (which you should identify), above which a further increase in pH results in a decrease in activity.

Explaining the trends

Your summary of the trends or patterns should help you to draw your conclusions. A good place to start is the hypothesis or the prediction that you were presented with initially. Do your results support or reject the hypothesis or prediction? Try to keep your conclusion concise.

Then, you need to add a paragraph or two to explain your findings using your biological knowledge. You should have revised the topic prior to undertaking the practical task, but be careful. You must be selective. Try to keep to information that is relevant to your results, and not to simply regurgitate information from textbooks.

The following are important when explaining trends in results:
- Have a clear picture in your mind of the relationship between the IV and the DV.
- Be clear about what your measure of the DV means. For example, if you have measured the time for an activity to be completed then you must realise that a smaller *time* represents a higher *rate* for the process taking place.
- Try to think why a change in the IV should change the DV.
- Keep what you write relevant but try to go into as much detail as possible. For example, with respect to enzyme activity, don't be content to just say that the enzyme is denatured. Provide further detail of protein structure and what bonds are being adversely affected.

Be careful that you do not simply describe the results, without identifying trends. This can happen when you quote values on a graph and ignore, for example, the phases of incline and/or plateau. Remember too that you should use *biological terminology* as fully and as carefully as possible.

When drawing conclusions, be careful not to suggest a causal relationship when one has not been proved. Ecological investigations often show a correlation between variables, but, because other variables are not controlled, cause and effect is not proven. Consider the investigation, the results of which are shown in Figure 29. A negative correlation between soil pH and thickness of the heather stem is apparent. It would appear that heather grows better in soils of lower pH, but it could just as readily be that the heather, or detritus from the heather, is causing the soil pH to decrease (become acidified); or it could be a combination of both. It is important to report the negative correlation and then discuss the possibilities, without making a firm conclusion.

You may also wish to draw attention to the biological significance of your findings.

Evaluation

Any practical or investigation will have limitations. To ensure the integrity of your conclusions it is important that you include an evaluation of the experiment that you have carried out. This should include an evaluation of the:

- limitations of the equipment
- limitations of the procedures
- reliability of the results
- validity of the conclusions

Evaluating equipment

When you come to evaluate the equipment that you are using, think about how precisely and accurately you were able to measure the DV.

Remember, *precision* is related to the smallest scale division on the measuring instrument that is being used. There are two things to consider when examining precision:

- Did you use equipment capable of the degree of precision required (e.g. use a syringe or measuring cylinder with $0.1\,cm^3$ gradations, rather than one with $1\,cm^3$ gradations)?
- Did you measure over a sufficient period of time (e.g. taking each potometer reading at 2-minute intervals rather than every 30 seconds; or leaving a catalase reaction for enough time to produce an appreciable volume of oxygen)?

You will also have to consider the accuracy of the instrument or apparatus that you used to make your measurements. Consider the apparatus that might be used for measuring oxygen production resulting from the action of catalase on hydrogen peroxide (see Figure 5 on page 19). There are alternatives: one that measures the amount of foam generated in the measuring cylinder, or one that collects the oxygen under water for measurement in a measuring cylinder (while there is also a gas syringe). Which will produce a more accurate measurement? Or, alternatively, which will produce a measure that is more prone to error? The level of foam rises in the measuring cylinder, but not always evenly and is also prone to quickly subside as the oxygen escapes. Similarly, in a photosynthesis experiment you could use apparatus that allows you to count the number of bubbles produced per unit time, or you could use the Audus apparatus to collect the volume of gas produced in a defined period and then measure it. Again, which apparatus will produce more accurate results? Try to reason this out.

Evaluating procedures

When evaluating procedures think about the following:

- How well were procedures standardised? For example, in the determination of water potential (of, say, potato tissue) a standardised drying technique is used to surface dry discs of tissue prior to weighing. However, this is not easy to achieve, particularly if you have not practised the procedure beforehand.

- How well were variables controlled? For example, you may have used a water bath to absorb heat from the lamp used to provide light for an aquatic plant (measuring gas production as a measure of photosynthesis). However, the water might still be expected to warm up, so you would need to monitor any increase in temperature periodically using a thermometer.
- How well were timings managed? For example, in the experiment on the course of an enzyme-controlled reaction, did you start the stopwatch as soon as the amylase was added to the solution of starch? Were you able to take a 1 cm³ sample immediately so that it represented a sample at time 0?
- How well were inconsistencies avoided? For example, in the determination of water potential experiment, did you remove a set of tissue discs and surface dry them before an electronic balance became available? If so, the discs would have been left exposed, water would have evaporated from the tissue and the mass would have been lower that it should have been. An inconsistency may show itself as an anomalous result.

Tip Human error occurs when you, the experimenter, make a mistake. Human errors are not a source of experimental error; rather, they are the experimenter's error. Avoid saying something like: 'I could have made a mistake in setting up the apparatus' or 'I made a mistake in taking a reading'. This not an evaluation — it is an admission of carelessness. If you could have been more careful, you should have been. So, do not quote human error as a source of experimental error.

Evaluating reliability

You can assess the reliability of your data in two ways:
- By comparing the spread of values for your replicates. You would look at your table of data to consider this.
- By looking at how your mean values fit a trend line. You can look at your graph of mean results to consider this.

Look at the results in Table 8. They are from an experiment into the effect of temperature on the digestion of starch by fungal amylase.

Table 8 The effect of temperature on the digestion of starch by fungal amylase (samples were taken after 2 minutes and measurements of remaining starch obtained using a colorimeter)

Temperature/°C	Colorimeter reading/% transmission			
	Replicate 1	Replicate 2	Replicate 3	Mean
20	87.6	68.8	95.0	83.8
30	44.6	39.2	48.8	44.2
40	27.1	25.2	12.8	21.7 [26.2]
50	11.4	9.4	13.1	11.3
60	48.2	46.4	50.0	48.2
70	98.2	95.2	100.0	97.8

At different temperatures, the amount of starch remaining after 2 minutes, following the addition of the enzyme, was measured using a colorimeter. You can see that the replicates for 20°C are quite variable (ranging from 68.8 to 95.0) and so the mean calculated might be regarded as less reliable. However, with one exception (discussed below), the trends are the same, and so you can be confident about making conclusions based on the mean values. You can also see that the replicates at the higher temperatures all have values that lie very close to each other. These results are very reliable. Note that as part of your *interpretation* you might suggest that the optimum temperature for fungal amylase is in the region of 50°C. Why can you not conclude that it *is* 50°C?

Inspection of the result highlighted (replicate 3 at 40°C) will indicate that it does not appear to fit the trend. This would also be obvious from a graph of the mean results along with those for replicate 3. The value may be judged as an anomaly. Since it does not appear to adversely affect the overall trend for mean values you might simply put up with it. However, if you can suggest a reason for the low value (12.8 in comparison with the others at 27.1 and 25.2) then you can justify ignoring it in calculating the mean (adjusted mean given in brackets). Is it possible that there was a drop in the electrical current just as the colorimeter reading was being taken? Could this have been due to electrical equipment being switched on in another laboratory?

It is important that you assess the reliability of *your results*. You must refer to *your data*. If they are found to be wanting then you must discuss where and how errors might have arisen. Avoid just saying that increasing the number of replicates would improve the reliability of the data. If a controlled variable was not kept constant during an experiment, then increasing the number of replicates would not improve reliability (and the validity of the experiment would be compromised).

Evaluating validity

Validity refers to the confidence that you can have in your conclusions. In a valid investigation:
- you have measured what you set out to measure
- changing the IV led to the changes in the DV that you have measured

You should already have discussed the following influences:
- the precision of the data collected
- the accuracy of the results
- the limitations in the procedure
- the effects of errors on the reliability of the results

Basically, you cannot make valid conclusions if there is serious doubt about the precision, accuracy and/or reliability of the results.

You should emphasise positive aspects of the investigation, followed by some criticisms. Refer to specific aspects of the procedure and results, rather than using vague comments such as 'my conclusion is valid because my results are precise, accurate and reliable' — this is meaningless without supporting information. For example, you could say that your results are precise because you measured to three significant figures and that they are reliable because the replicates are close together and there are no anomalous data. Your results may be accurate because they agree with the expected trend. If you can identify possible errors then comment on them and suggest what might have been done to minimise their influence.

Another aspect of validity is the degree to which conclusions reached about relationships between variables are justified. This includes any assumptions that may have been made. For example, a potometer is used to investigate transpirational loss of water, but actually it is measuring water uptake. Not all water taken up is lost by transpiration; some is used to maintain turgidity and, in leaves, may be used in photosynthesis. You would need to note this and reason that, as other limiting factors are minimal, any water lost is replaced by that taken up.

In ecological work there are so many factors operating that it is difficult to establish causal relationships. Look at Figure 32. This shows the results of an investigation into the distribution of mayfly larvae and the speed of the water current in a river. (A scatter graph has been plotted with 'speed of current' along the *x*-axis — it seems reasonable to assume that the speed of the water current is not dependent on the number of mayfly larvae present.) It is not possible to state a causal relationship. It is more likely that water current is linked with some other factor, which has a more direct bearing on the numbers of mayfly larvae present. You might argue that oxygen levels in the water would be higher where the current is stronger; and that mayfly larvae have a niche requirement for high oxygen levels.

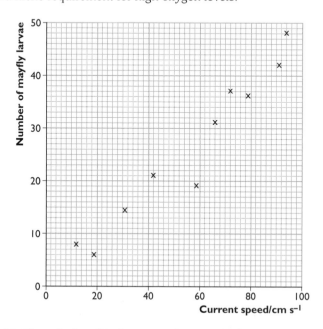

Figure 32 The relationship between the speed of water current and the number of mayfly larvae found in sections of a river

Assessment of practical tasks at AS: exemplars and comments

At AS, you will be assessed over *two practical tasks*. It is important that the practical tasks used for assessment are of sufficient rigour to allow the demonstration of practical skills appropriate for AS.

In each practical task you will be assessed over four **skill areas**, one of which is subdivided:

- **Implementing** a sequence of instructions
- **Recording and communicating** results
  - **Tabulation** of raw and derived data
  - **Graphical** presentation of the data
- **Interpretation** of the results
- **Evaluation** of the experimental design

Each of these areas has **five skills** (with mark descriptors), and will be marked by your teacher on a three-point scale: 0, 1 or 2 marks.

0 marks will be awarded under the following circumstances:

- There is no evidence of the skill having been demonstrated, e.g. there is no caption on the graph.
- The attempt to demonstrate the skill is not appropriate, e.g. neither 'my results' nor 'table of results' is an appropriate caption for a graph.
- The attempt to demonstrate the skill lacks evidence, e.g. simply saying, in an evaluation of results, that 'my results are valid' without offering any supporting evidence.
- The attempt to demonstrate the skill offers only an incorrect answer, e.g. 'a high % transmission indicates a lot of pigment present'.
- The attempt to demonstrate the skill lacks relevance, e.g. in an experiment on the effect of pH on enzyme activity, only discussing the effect of temperature.

2 marks will be awarded for the demonstration of a high level of competence in a skill. **1 mark** is awarded where there is a failure to demonstrate full competence in a skill.

The assessment of the skill areas is illustrated below by the use of two practical tasks, each of which is undertaken by a student. The first practical task, *The effect of temperature on membrane permeability in beetroot*, is used to illustrate the assessment of **skill area A** (Implementing), **skill area B1** (Tabulation of data) and **skill area D** (Evaluation of the experimental procedure). The second task, *Investigating the water potential of potato and sweet potato*, is used to illustrate the assessment of **skill area B2** (Graphical presentation of data) and **skill area C** (Interpretation).

Practical task 1: the effect of temperature on membrane permeability in beetroot

In beetroot cells the red betalain pigment occurs in their sap vacuoles. Consequently betalain is isolated from the exterior of the cell by several layers, including two membranes. This experiment investigates how temperature affects the permeability of these layers, consequently influencing the quantity of betalain pigment leaving the cells. A suitable hypothesis would be: *At a certain high temperature the membranes become permeable to betalain, causing it to leak out of beetroot cells.*

A summary of the procedure presented to students is shown below.
- Using a Bunsen burner, heat 200 cm³ of distilled water to 85°C.
- While the water is heating, use a pipette to add 10 cm³ of unheated distilled water to each of nine test tubes. Label the tubes 85, 80, 75, 70, 65, 60, 55, 50 and 45 respectively.
- Use a cork borer to obtain a beetroot cylinder and cut this to 5 cm in length. Place it in the beaker of water at 85°C, and note the time. After exactly 1 minute use forceps to transfer the cylinder to the test tube of cold water labelled 85. Note the time.
- As the water in the beaker cools, repeat the step above, using a fresh cylinder each time, at the following temperatures: 80, 75, 70, 65, 60, 55, 50 and 45°C.
- Leave each cylinder in its test tube of cold distilled water for exactly 30 minutes, and then remove it.
- Using a colorimeter, measure the percentage transmission of the pigment that has escaped from the cylinders at each treatment temperature.

Student A's responses in skill areas A, B1 and D

Skill area A (Implementing)

The implementation of the procedure was observed and marked by the teacher.

✐ Student A scores 9 marks as follows:

Skill	Examiner's comment	Mark
Skilful handling of apparatus and materials	The setting up of a water bath, preparation of beetroot cylinders and use of the colorimeter were all well carried out.	2
Adherence to appropriate safety procedures, etc.	The student wore a lab coat while doing practical work and cut cylinders of beetroot carefully.	2
Organised and methodical carrying out of procedure	The student kept a tidy, well organised space on the bench and worked methodically through the procedure.	2

Skill	Examiner's comment	Mark
Adhering to the instructed sequence of actions	Most actions were in sequence though the student had a beetroot cylinder prepared some time before the beaker of water was heated to 85°C.	1
Measurements at the level of precision required	The student was able to obtain a colorimeter reading (% transmission) accurately to one decimal place and, more importantly, to three significant figures (using the digital read-out available).	2

Skill area B1 (Tabulation of data)

A table of results for replicates and means

Temperature	Replicates					Mean
	1*	2	3	4	5	
45°C	45.1	93.6	96.7	96.7	98.2	86.1
50°C	88.4	93.4	40.5	91.2	95.8	81.9
55°C	83.4	14.6	84.3	88.4	19.2	58.0
60°C	78.6	10.5	71.6	58.5	12.7	46.4
65°C	45.9	1.2	39.8	1.8	6.9	19.1
70°C	4.7	0.6	13.6	3.5	1.2	4.7
75°C	7.2	0.8	1.7	3.3	4.3	3.5
80°C	0.8	4.3	3.9	1.7	5.0	3.1
85°C	0.6	1.2	0.4	3.4	8.9	2.9

*Student's own results — these were presented in a separate table for marking (and included the same weaknesses as illustrated in this table).

Student A scores 6 marks as follows:

Skill	Examiner's comment	Mark
Organisation of the data for independent and dependent variables in the table	The table shows the independent variable in the first column and the dependent variable in the body of the table.	2
Caption including the independent and dependent variables and the material used	The caption to the table lacks any detail about the variables or the material used.	0
Logical construction of table — to facilitate analysis	The construction of the table readily facilitates analysis, e.g. the variation in the replicates can be easily assessed while the amount of pigment escaping at different temperatures can be readily interpreted.	2
Explanatory column headings	The student's table shows the column heading for temperature (the independent variable) but lacks the heading for percentage transmission (the dependent variable).	1

Skill	Examiner's comment	Mark
Units of measurement for all variables and calculated values	With respect to the units of measurement, the student only scores 1 mark since '°C' is shown in the body of the table rather than with the column heading.	1

Try this yourself

You will benefit from completing the construction of a graph and an interpretation of the results.

For the graph consider the following:
- the most appropriate form of graph
- the most appropriate caption
- the labels and units of measurement to be presented on the axes
- ensuring that the IV is on the *x*-axis
- appropriate scaling of the graph
- accurate plotting of points
- joining points either with straight lines or a line of best fit. While joining points with straight lines is easier to accomplish, you should practise drawing lines of best fit according to the rules (see page 51).

In your interpretation you should:
- describe the results, noting that a high % transmission indicates that there is little pigment present while a low % transmission means that a lot of pigment has leaked out
- identify the trends in the graphs by asking yourself if the rate of pigment release is constant as temperature increases
- attempt to explain the findings using your understanding of:
  - plant cell structure (if the betalain is in the sap vacuole then how many layers does it have to pass through in order to escape out of the cell)
  - membrane structure
  - the effect of temperature on the components of the cell membrane

Remember to use the appropriate terms such as 'tonoplast' for the membrane of the sap vacuole or increased 'fluidity' of the phospholipids or 'denaturation' of transmembranal proteins when considering the effect of temperature on membrane structure.

Skill area D (Evaluation of the experimental procedure)

All measurements were accurate. Volumes of water were measured out using a $10\,cm^3$ pipette. The temperature of the water was measured using a thermometer, and the range of 45°C to 85°C was appropriate simply because it included the temperature at which pigment leaked out of the cells. Pigment density was determined using a digital colorimeter and the % transmission reading recorded to 1 decimal place.

The control of variables ensured a fair test. This included: same size of beetroot cylinder used; same volume of cold distilled water into which heated cylinders were added; and each cylinder was left in the water for exactly 30 minutes, timed with a stopwatch.

Since the measurements were accurate then the results were valid.

The results (see table) are variable, especially at the lower temperatures. For example, at 65°C the range is 1.2 to 45.9, at 60°C it is 10.5 to 78.6, at 55°C it is 14.6 to 88.4, at 50°C it is 40.5 to 95.8 and at 45°C it is 45.1 to 98.2.

Where the range is wide the results must be regarded as unreliable. So it appears that the results at higher temperatures (70°C to 85°C) are reliable while those at lower temperatures (from 65°C to 45°C) are less reliable.

Student A scores 6 marks, as follows:

Skill	Examiner's comment	Mark
Appropriateness of the measurements and observations	The candidate has completed appropriate comments with respect to the accuracy of measurements and concerning the temperature range used.	2
Procedures used for the control of variables not under investigation	There is also a sound understanding of how factors not under investigation are controlled.	2
Reference to the validity of the experimental design and procedures	However, there is no understanding of validity within the experiment and so for this skill no marks are awarded. Does the colorimeter reading provide a valid measure of the amount of pigment that has leaked out of the beetroot cells? When asked about validity, you are asking yourself the question, 'am I measuring what I think I am measuring?'	0
Assessment of the variation shown by the replicates	The student has attempted to assess the variation among the replicates for each treatment but, unfortunately, refers only to the range. The range compares two values, the highest and the lowest, rather than the 'closeness' of the majority of values, so only 1 mark is awarded.	1
Comments on the reliability of the data	Again, when commenting on the reliability of the data the student shows some understanding of the relationship between the range width and reliability but has ignored how well the mean values fit a 'smooth' overall trend, so 1 mark only is awarded.	1

Practical task 2: investigating the water potential of potato and sweet potato

Plant storage organs, such as potato tubers and sweet potato tubers, contain stores of carbohydrates, which may be soluble sugars or insoluble starch. Potatoes mostly store their carbohydrate as starch, while sweet potatoes contain stored sugars. Solutions with a higher concentration of solutes tend to have a lower water potential. Consequently, a hypothesis may be formulated: *Sweet potato tuber tissue has a lower average water potential than potato tuber tissue.*

A summary of the procedure presented to candidates is shown below.

- Label six specimen tubes: 0, 0.2, 0.4, 0.6, 0.8 and 1.0 mol dm^{-3}. Place approximately one third of a tube of distilled water in the first, and an equal volume of each of a series of sucrose solutions of different strengths (molarities) in the remainder. Each tube should be firmly stoppered.
- Using a cork borer and a razor blade, prepare six solid cylinders of potato. Each cylinder should be approximately 10 mm diameter and 12 mm long. Slice up each cylinder into six discs of approximately equal thickness. Place each group of discs on a separate piece of filter paper. Turn once to remove surplus fluid from both faces of the discs.
- Weigh each group of discs. (In each case weigh them on a piece of filter paper, then weigh the filter paper alone, and subtract the one from the other to get the mass of the discs.) Record the mass of each group. This is the initial mass (g).
- Put one group of discs into each of the labelled tubes immediately it is weighed. Stopper each tube firmly and leave for 24 hours.
- After 24 hours remove the discs from each tube. Remove any surplus fluid from them quickly and gently with filter paper, using a standardised procedure for all of them. Then re-weigh them. Record the new mass of each group of discs. This is the final mass (g).
- Present the results in an appropriate table showing the molarity of each sucrose solution, its water potential (using the calibration table available), and the percentage change in mass (change in mass multiplied by 100 divided by the initial mass).
- Repeat the above steps for sweet potato tuber tissue.
- Plot a graph of percentage change in mass for each of the two plant tissues against water potential of the immersing sucrose solution, and use this to calculate the water potential of each tissue.

Student A's responses in skill areas B2 and C

Skill area B2 (Graphical presentation of data)

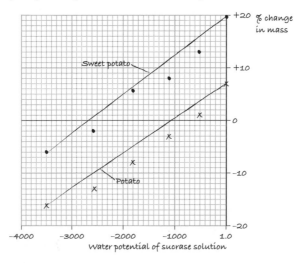

The percentage change in mass of potato and sweet potato

Student A scores 7 marks, as follows:

Skill	Examiner's comment	Mark
Selection of appropriate form of graph	The student has selected the appropriate line graph.	2
Caption including the independent and dependent variables and the material used	While there is a caption, the independent variable (the different sucrose solutions) has been ignored and so only 1 mark is awarded.	1
Data plotted correctly with appropriate scales	The data have been accurately plotted.	2
Labels for both axes including units of measurement	The axes have been appropriately labelled though the student has omitted the unit of measurement for water potential of sucrose solution (kPa) and receives only 1 mark.	1
Appropriate lines drawn with key as appropriate	The student has labelled each line, which is acceptable (alternatively a key to the lines might have been presented). Lines of best fit have been attempted and this is appropriate since calculations have to be made. However, the student's lines are not the best fit and would appear to be connecting just the top and bottom points, so only 1 mark is awarded.	1

Skill area C (Interpretation)

From the graph, I can see that potato and sweet potato act in similar ways. Both have increased in mass in sucrose solutions with a high water potential (dilute solutions) and lost mass in solutions with a very negative water potential (concentrated solutions). Although they are similar, the lines of best fit cut the *x*-axis at different points indicating different equilibria with the sucrose solution.

The water potential of a tissue can be determined where the line of best fit crosses the x-axis. For potato this is −1080 kPa and for sweet potato it is −2680 kPa. Sweet potato has a more negative water potential. This agrees with the hypothesis that 'sweet potato will have a lower water potential than potato'.

Water molecules move from a high concentration to a low concentration (?)*. So if a solution has a higher water concentration than the plant tissue, then water will move into the tissue and it will increase in mass. The reverse will happen if the solution has a lower water concentration. Where there is no change in mass then the tissue and the solution may be said to have an isotonic concentration.

In potato starch is the main storage carbohydrate and since it is insoluble it is osmotically active (**✗**)*. This is why more water is drawn out of the potato over a wide range of solutions. In sweet potato, the sugars are soluble and so exert an osmotic effect. This means that more water is drawn into the cells, increasing the overall mass of the tissue.

* *teacher's annotations*

▨ Student A scores 7 marks as follows:

Skill	Examiner's comment	Mark
Written communication of the data	The student communicates an understanding of the relationship between the gain or loss in mass of tissue with the water potential of the external solution.	2
Trend(s) clearly identified	There is understanding of how to determine the water potential of each tissue and these have been correctly determined.	2
Explanation of the trend(s)	However, there are limitations when it comes to offering an explanation. The student refers to water concentration rather than water potential, and could be clearer in explaining the equilibrium situation between the tissue and the external solution — that when there is no net change in mass the water potential of the two systems is equal. Only 1 mark is awarded.	1
Use of appropriate biological understanding	When referring to the storage carbohydrate the student makes a mistake and should have described starch as 'osmotically *inactive*'. However, there is some understanding that sugars will 'exert an osmotic effect' and so 1 mark is awarded.	1.
Use of appropriate terminology	Since there is limited use of the appropriate biological terminology only 1 mark is awarded for this skill.	1

Try this yourself

You will benefit from completing an evaluation of the experiment.

While you cannot comment on an assessment of the variability of the replicates and of the reliability of the results (since tables of the data are not presented) you should consider the following:
- the appropriateness of the range of sucrose solutions $(0-1.0 \, mol \, dm^{-3})$
- the accuracy of the measurements, for example the masses of potato discs
- the control of other variables
- aspects of validity
  - Is the change in mass entirely due to osmotic movement of water in or out of the plant tissue cells?
  - Do different cells within a tuber all have the same water potential?
- the effectiveness of a standardised drying technique and the effect on the results
- the effect of taking cylinders of tissue from different parts of a potato (or sweet potato) tuber

Investigational and Practical Skills in A2 Biology

This section is divided into three parts:

- **Further practical work at A2** — this part covers all the practical work that you have to carry out at A2 and includes practicals on populations, respiration and photosynthesis:

Populations (A2 1) ....page 71

Respiration (A2 2)..page 73

Photosynthesis (A2 2) ....page 75

- **Demonstrating investigational and practical skills at A2** — this part covers all the skills that you have to demonstrate when assessed on the investigation. These include:

Planning: developing a hypothesis..page 77

Planning: planning a procedure ....page 77

Planning: planning for statistical analysis .... page 83

(with a section on *Statistical techniques*)..pages 79–83

Implementation ....page 85

Recording data in a table..page 85

Statistical analysis ....page 85

Interpretation ....page 88

Evaluation..page 89

- **Assessment of a practical investigation: exemplars and comments** — this part shows the skill areas within which you are assessed. Two exemplar investigations are presented:

Investigation 1: the effect of biotin on the growth of a yeast population (assessing *Developing a hypothesis, Planning a procedure, Planning for statistical analysis* and *Implementation*) ....page 91

Investigation 2: investigating resistance to desiccation of three species of rocky shore seaweeds (assessing *Tabulation of data, Statistical analysis, Interpretation* and *Evaluation*)..page 95

Further practical work at A2

Populations (A2 1)

Investigating the growth of a yeast population using a haemocytometer

The population of a yeast culture can be investigated by adding a sample of yeast suspension to a flask of nutrient medium, which has been previously sterilised. A suitable nutrient medium for yeast is a 2% sucrose solution.

The flask containing the yeast cells and the nutrient medium is plugged and kept in an incubator at constant temperature (say, 25°C) over several days. Samples are taken at intervals (at least twice a day) and the yeast cells counted using a haemocytometer. Cell density (cells mm^{-3}) is plotted against time (hours).

A haemocytometer consists of a special glass slide with an accurately ruled, etched grid of precise dimensions on a central, lowered platform flanked by grooves (see Figure 33). The small, Type-C square is 0.05 mm × 0.05 mm = 0.0025 mm^2 in area. When the coverslip is correctly positioned over the counting grid, the depth of the counting chamber is 0.1 mm, and the volume of a Type-C square is therefore 0.00025 mm^3 (or 1/4000 mm^3).

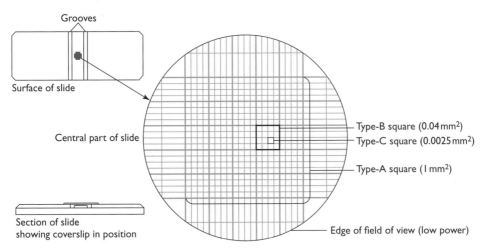

Figure 33 A haemocytometer

To obtain a cell count, a representative sample of the yeast culture is placed beside the coverslip so that capillarity draws it under. The number of cells in 40 Type-C squares is counted and totalled. Different numbers of squares may be counted, but 40 facilitates the calculation of cell density using the formula:

$$\text{number of cells per mm}^3 = \frac{\text{total number of cells in 40 Type-C squares}}{40} \times 4000$$

The graph should show a phase of exponential increase followed by saturation (when the curve levels or plateaus) and may even include a decline phase (depending on the length of time during which samples are taken). An explanation of the exponential phase should indicate that resources are plentiful and not limiting yeast cell division. A plateau phase occurs when some factor is becoming limiting — such as lack of resources or accumulation of waste (ethanol or carbon dioxide, which would cause the pH to fall). A decline phase occurs when many cells are dying (due to lack of resources or accumulation of waste) and not being replaced.

There are many factors to be careful about when doing this experiment:
- A representative sample of the yeast culture is obtained when the culture is shaken to ensure a homogeneous suspension before a sample is removed.
- The sample on the haemocytometer must not enter the grooves, otherwise the procedure has to be repeated.
- When counting in the Type-C squares include those cells touching the top and left boundaries (sometimes called the north-west rule) and exclude those touching the right and bottom. Applied consistently, this ensures that no cell can be counted twice.
- Samples taken when the yeast population has become dense may mean that there are too many cells to count. In this situation, an accurate dilution, say 1 in 10, should be made and the final count should then be multiplied by the dilution factor.
- The haemocytometer gives a total count including both living (or viable) and dead (non-viable) cells. Living and dead cells can be distinguished by the use of special staining techniques — if methylene blue is added to the sample, dead cells are identified as being blue, while living cells are colourless (see page 74 for why this is the case).

Question 1
The average number of cells in a Type-C square was determined as 8.15. Calculate the number of cells per mm^3.

Estimating the size of an animal population using a simple capture–recapture technique

Many animals move and often, during daytime, remain hidden. The technique for estimating their population size is called **capture-recapture** (also called **mark-release-recapture**). A sample of individuals is caught, counted and marked in some way — this is the first sample (s_1). These marked individuals are released back into their original location. After being allowed to mix with the unmarked individuals in the population, a second sample (s_2) is caught and counted and the number of marked individuals noted — these are the recaptures (r) since they were also caught on the first occasion. An estimate of the total population size can then be made by calculating what is called the **Lincoln index** (also called the **Peterson estimate**) as follows:

$$\text{population size} = \frac{s_1 \times s_2}{r}$$

The method of capture and marking needs to be appropriate for the animal species being investigated. For example, ground beetles can be caught in pitfall traps (see page 34) and marked by placing a minute drop of correction fluid or quick-drying waterproof paint on one of their hardened front wings.

Using this technique for estimating population size relies on a number of assumptions:
- The mark should not harm the animal (this can be tested by keeping a sample of marked individuals in the lab to check for toxicity), it should persist over the sampling period, and it should not make the animal more obvious to predators or influence its behaviour. If these requirements are met, the probability of capturing a marked individual should be the same as that of capturing any member of the population.
- The marked individuals are completely mixed in the population, and enough time has elapsed between visits to the study area to allow this to happen.
- The population is 'closed', that is, the two visits to the study area are close enough in time so that no individuals die, are born, move into the area (immigrate) or move out of the area (emigrate) between visits. If the above requirement is met, it may be assumed that the proportion of marked individuals in the second sample is equivalent to the proportion of marked individuals in the whole population.
- Between 10% and 30% of the population (as measured by the ratio of marked to unmarked individuals caught in the second sample) should carry marks if the estimate is to be reliable.
- If the population is an 'open' one then:
  – if there are losses from the population (through deaths or emigration) during the study period, the estimate is for the size of the population at the time of the first sampling session
  – if there are gains (through births or immigration), the estimate corresponds to population size during the second sampling session

Question 2
A sample of 50 snails was removed from a study site, marked and released back into the population. Three days later a second sample of snails was collected and of these 48 were unmarked and 12 marked. Calculate the estimate of the population size.

Respiration (A2 2)

Measuring the respiratory quotient using a respirometer

The **respiratory quotient** (**RQ**) is a measure of the ratio of carbon dioxide given out by an organism to the oxygen consumed over a given time period:

$$RQ = \frac{\text{volume of } CO_2 \text{ given out}}{\text{volume of } O_2 \text{ taken in}}$$

The volume of O_2 taken in by living organisms, such as bean seeds, is determined using a respirometer with potassium hydroxide to absorb the CO_2 given out (see page 27). The volume of CO_2 given out is determined, for the same bean seeds, using a respirometer with water replacing the potassium hydroxide. This measures the net difference between CO_2 production and O_2 consumption and so, having already determined the volume of O_2 taken in, the volume of CO_2 given out can be calculated.

RQ values provide clues to the substrate being respired and the extent to which the tissue is respiring anaerobically (see the Student Unit Guide for A2 Unit 2, page 17).

Question 3
A respiratory quotient of 0.9 was obtained for some seeds. Discuss.

Use of redox indicators to demonstrate dehydrogenase activity in respiration

During respiration, particularly during the Krebs cycle, dehydrogenase enzymes remove hydrogen, which is taken up by hydrogen acceptors (NAD^+ generally, but also FAD) that subsequently become reduced. Experiments can be devised in which the hydrogen released by dehydrogenase activity in living yeast cells is taken up by artificial hydrogen acceptors, which change colour when reduced. Since the 'artificial' acceptors are one colour when oxidised (the normal state in an atmosphere containing oxygen) and another colour when reduced, they are called **redox indicators**. Three redox indicators are shown in Table 9.

Table 9 Some redox indicators and their colours when oxidised and reduced

Redox indicator	Colour when oxidised	Colour when reduced
Methylene blue	Blue	Colourless
Triphenyltetrazolium chloride (TTC)	Colourless	Pink
Dichlorophenol indophenol (DCPIP)	Blue	Colourless

Dehydrogenase activity in respiration can be investigated using a range of living materials, e.g. seeds and yeast. A redox indicator is added to a suspension of the living material and observations made of the colour change as the indicator is reduced. For example, using methylene blue, a measure can be made of how long it takes for the blue colour to disappear. A colorimeter could be used as a quantitative means of measuring the reduction of the indicator — percentage transmission should increase as the solution loses its blueness.

Redox indicators can be used in various experiments on dehydrogenase activity:
- a demonstration of hydrogen release in a living yeast suspension compared with a control containing 'boiled' yeast
- the effect on dehydrogenase activity of adding an intermediate of the Krebs cycle, such as succinate (which is the substrate for succinate dehydrogenase)
- the effect of malonate (a competitive inhibitor of succinate dehydrogenase) on dehydrogenase activity

Question 4
During an experiment on the respiration of yeast, using methylene blue as a redox indicator, students were directed to ignore the thin film of blue colour at the surface. Suggest why.

Photosynthesis (A2 2)

Using the Audus apparatus to investigate the effect of factors on the rate of photosynthesis

The use of the Audus apparatus to investigate the effect of light intensity on the rate of oxygen production is described on pages 28–29. The experiment can be modified to measure the effect of other factors on the rate of photosynthesis: wavelength (colour) of light, temperature and carbon dioxide.

Question 5

Suggest how you would vary each of these factors to investigate their individual effects on the rate of photosynthesis: wavelength of light, temperature and carbon dioxide.

Carrying out paper chromatography of plant pigments

Photosynthetic pigments can be separated and identified by paper chromatography (see page 14). The pigments are extracted from a leaf and a concentrated spot created on the chromatography paper. The different pigments are identified by their different colours and positions (see Figure 34). R_f values are calculated by adapting the formula on page 14.

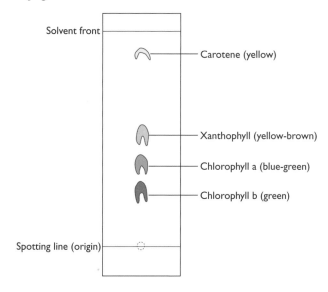

Figure 34 Plant pigments separated on a paper chromatogram

The role of different pigments in absorbing light of different wavelengths may be discussed. It is also possible to analyse the pigments in different seaweeds.

Question 6

Explain the role of different pigments in photosynthesis.

investigational and practical skills in A2 Biology

Demonstrate the role of hydrogen acceptors using a redox indicator (such as DCPIP)

During the light-dependent stage of photosynthesis, electrons are energised to reduce NADP$^+$. (The NADPH is used subsequently in the light-independent stage of photosynthesis.) Using DCPIP to accept electrons demonstrates the activity of the light-dependent stage experimentally. As the DCPIP is reduced it turns from blue to colourless.

Demonstration of the activity of the light-dependent stage of photosynthesis requires the following:
- Isolated chloroplasts. These are obtained by grinding leaves, filtering to remove debris and centrifuging the filtrate so that chloroplasts are thrown down into the precipitate. (It is important to keep the suspension and apparatus cold during the extraction procedure.)
- A suitable redox indicator. Electrons will be accepted by the blue dye DCPIP, which when reduced turns colourless.

DCPIP is added to a suspension of chloroplasts (a control would involve DCPIP added to water), and the time it takes to decolourise the DCPIP measured. The loss of colour in the DCPIP is due to electrons produced by light-dependent reactions in the isolated chloroplasts. A control in which DCPIP is added to water would remain blue.

Question 7
DCPIP added to a suspension of whole plant cells turned from blue to colourless. Explain why this does not demonstrate solely the activity of the light-dependent stage of photosynthesis.

Demonstrating investigational and practical skills at A2

At A2, you will be assessed on a *Practical Investigation* which must include the following skill areas:
- **Planning** the investigation
  - **Developing a hypothesis** using biological knowledge
  - **Planning a procedure** to test the hypothesis
  - **Planning for statistical analysis**
- **Implementing** a procedure
- **Tabulation** of data
- **Statistical analysis** of the data
- **Interpretation** of the results
- **Evaluation** of the practical procedures

This section will deal mostly with those aspects of investigative skills that are unique to A2 — that is, planning and statistical analysis.

Planning

While you may be provided with a topic for investigation, it will be up to you to plan what you do. There are three stages to this:
- **Developing a hypothesis**
- **Planning a procedure**
- **Planning for statistical analysis**

Developing a hypothesis

Having been provided with a topic for investigation, you will need to undertake **research**. The first place to start is your course notes and textbooks. Other sources include biology books in the library and websites on the Internet. While you may read around the topic, it is important when it comes to writing up your investigation that you are *concise and relevant*. You must avoid writing everything you know about the topic. You must be selective. For example, if you are investigating the effect of pH on fungal amylase, then you must outline your biological knowledge on protein structure (since enzymes are proteins), the effect of pH on structure (but not the effect of temperature) and you may discuss some aspects of fungal enzymes, such as the type of environment that they work in.

You must produce a clear, well-defined, testable **hypothesis**. A hypothesis is essentially an 'educated guess'. For example, you might state that a fungal enzyme will have an optimum pH at which it works best and that its functionality will decrease at pHs below or above this optimum. What is not acceptable, however, is simply to say that pH affects fungal enzyme activity. The hypothesis must be directional — it must anticipate the trend.

Planning a procedure

Your hypothesis must be testable and you must suggest a procedure by which you can investigate your idea. For example, if you wish to investigate the effect of pH on fungal amylase then you have to design a suitable experiment. Obviously, you are going to investigate the effect of fungal amylase on a solution of starch, but how are you going to measure starch digestion. You might use 'time for starch to be digested', testing samples at 30 second intervals with dilute iodine (see page 18). But would this provide the level of precision required? There are always problems in deciding when a blue-black colour is no longer produced. So, you might decide to use a colorimeter and test a sample after digestion for a suitable period of time, say 1 or 2 minutes.

Deciding what you are going to measure will provide you with your **dependent variable** (DV). You must also *decide on a suitable range* for the **independent variable** (IV). Do you want to study the effect over a wide range or focus on a narrow range? If you obtained information during your research that the optimum pH for fungal amylase is in the region of pH 6, then you might select a range of pH 4 to 8; otherwise, you would chose a reasonably wide range within which the optimum should be included (say, pH 2 to 10). You will also have to decide on the number of pH values to use. You *need to produce enough data points to allow you to identify a trend*. The minimum number is five and, while more is better, you need to consider the total amount of work involved. You can then see which pH buffer solutions are available to you. Do not worry if the available buffers are not whole numbers, e.g. pH 3, pH 4, etc. There is nothing wrong with using buffers at ph 3.1, 3.9, 5.2, etc. as long as they cover a suitable range — the pH scale is a continuous variable.

You will also now be able to make a **prediction** based on your hypothesis and what you are going to measure. You know that a low level of starch in solution will occur when a lot of digestion has taken place and that this will give a high % transmission reading on the colorimeter. So your prediction would be that the highest % transmission reading (after a period of digestion) would indicate the region of optimum pH — it cannot indicate this precisely unless you use a much greater number of pH buffers in this region.

Of course, while pH will influence the action of fungal amylase, so will many other factors. You should make a list of all the variables that could affect your investigation. All the other variables, such as temperature, enzyme and substrate concentrations and volumes, have to be controlled in order to produce valid results. They are the **controlled variables**. Your plan must include some information on how you intend to control these variables. Some will be relatively straightforward, for example use the same concentration and volume of starch solution throughout the investigation. However, others will require more careful management. You cannot assume that the variables will stay constant, so you will need to show how you will monitor them during the investigation. For example, when using a water bath to control temperature you should realise that the water temperature may fluctuate, so you would need to take regular temperature readings. If the temperature changes, you need to consider corrective action to bring the temperature back to the specified level.

Many investigations require the use of a **control experiment**. Controlled variables should not be confused with a control experiment. A control experiment is a comparative experiment that you set up to eliminate certain possibilities. In the enzyme investigation described above, a suitable control could be one treatment in which the enzyme is replaced by distilled water. Not all investigations can use a control, but it is important to include one where appropriate. Every plan should have a comment about the use of a control, even if it is just to state that it is not possible or necessary to have a control. In the investigation described, you have comparisons with the set-ups at different pHs, so a control is not really necessary.

Having designed a plan for the investigation where you have a designated IV, over a suitable range, decided on the DV you are going to measure and considered those variables that you need to control, you should then list all the equipment and pieces of apparatus that you will need to carry out the investigation. Measuring instruments, such as measuring cylinders, should be chosen to allow you to carry out measurements to an appropriate level of accuracy and precision. For example, if you are planning to take 1 cm³ samples of the enzyme–substrate mixture, then a 1 cm³ pipette or syringe would be appropriate.

Finally, you should write out a procedure for the investigation. This must indicate the sequence that you intend to follow. An ordered sequence might well take the form of a flow diagram, which provides a visual representation of what you are going to do.

Statistical techniques

In A2 investigations, you must plan how you are going to analyse the results. This must involve a statistical analysis of the data. The method of analysis must be appropriate for the results that you intend to record. Before you decide on the method of analysis ensure that you understand the statistical techniques available.

Need for statistical analysis

It has already been established that biological data are highly prone to variation. Random errors or variability in the biological material will cause data to deviate from an expected or true value. Statistics is the use of mathematical methods to describe data and to determine the probability of events, such as whether differences are due to random factors.

The statistical technique that is appropriate will depend upon what you are dealing with:

- **Measured data** — where you have obtained results by measuring something (say, rate of reaction), repeated the results and calculated a **sample mean**. Note that even if you repeated results in a laboratory a number of times (e.g. five repeats for each pH in the investigation of fungal amylase activity) the replicates still represent a sample of what you could potentially have done, say hundreds of repeats; and the calculated mean represents a sample mean.
- **Frequency data** — where you have made **counts** of something, say the number of germinated seeds out of a total number used, or the number of woodlice in either side of a choice chamber, one side of which is dark and the other light.

Statistical analysis of sample means

The number of measurements that can be made is limited (for example, by time constraints or availability of equipment) and so the measurements represent a sample. All the measurements that might be made represent the population (whether real, such as a population of wild garlic plants, or imaginary, such as all the measurements of amylase activity that could potentially be taken). The sample provides an estimate for the population. There are two potentially important properties that summarise the sample of data collected:

- The **sample mean** (symbol $\bar{x}$) — a measure of central tendency or average. (There are other measures of central tendency: the mode — the most numerous value; and the median — the value midway between the highest and the lowest. But the mean is the most helpful.)

investigational and practical skills in A2 Biology

- The **standard deviation** — a measure of the variability (or spread or dispersion) of the data. Where the variability of the population is estimated, this is denoted by the symbol $\hat{\sigma}$.

A sample mean ($\bar{x}$) provides an estimate of the mean of the population from which the sample has been drawn. The population mean is referred to as the true mean (and is denoted by μ). How reliable is the sample mean, that is, how close does it lie to the true mean? The reliability depends on two things:

- the sample size, n
- The variability of the data, as measured by $\hat{\sigma}$ (the estimate of the standard deviation)

Since sample means vary, it is important to consider how good an estimate any one sample mean might be. A sample mean will be more likely to lie close to the true mean if the sample size is large and the standard deviation is small. This is measured in a statistic called the **standard deviation of the mean** (also called **standard error of the mean**) with the symbol $\hat{\sigma}_{\bar{x}}$:

$$\hat{\sigma}_{\bar{x}} = \sqrt{\frac{\hat{\sigma}^2}{n}}$$

The standard deviation of the mean is a measure of how much the sample means would on average differ from the population mean. A small $\hat{\sigma}_{\bar{x}}$ value, relative to the magnitude of the sample mean, indicates a reliable sample mean — that is, one that that lies close to the population mean.

95% confidence limits

Even better is to estimate the boundaries within which the true mean might lie. Since it is not possible to have absolute certainty, a convention of 95% probability has generally been accepted in biology. 95% confidence limits are provided by:

$$\bar{x} \pm t(\hat{\sigma}_{\bar{x}})$$

where t is determined from a table of t values at $p = 0.05$ and $n - 1$ degrees of freedom.

95% confidence limits are also plotted on graphs. Figure 35 shows a bar graph for the mean height of pea seedlings in the light and in the dark. The 95% confidence limits are also shown. These limits set the boundaries within which there is a 95% probability of the true mean lying. You can see that, in this example, the limits for the results in the light and the dark *do not overlap*. This suggests that samples are significantly different — the difference between the sample means is real and not due to chance. This decision can only truly be made by undertaking a t-test. However, using 95% confidence limits is the only way in which you can make decisions about significant differences when you are comparing more than two sample means.

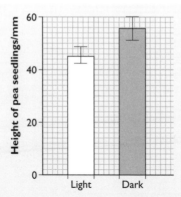

Figure 35 A comparison of the height of pea seedlings grown in the light and in the dark — means and 95% confidence limits plotted

t-*test*

The *t*-test (sometimes called Student's *t*-test) is a strong statistical procedure for comparing two sample means. Remember that sample means are expected to differ. The *t*-test allows you to determine if the difference is significant, that is, not just due to random factors or chance. The formula for calculating *t*, in terms of $\hat{\sigma}_{\bar{x}}$, is given as:

$$t = \frac{\bar{x}_1 - \bar{x}_2}{\sqrt{\hat{\sigma}_{\bar{x}_1}^2 + \hat{\sigma}_{\bar{x}_2}^2}}$$

A starting point in statistical tests is to establish a **null hypothesis** (given the symbol $\mathbf{H_0}$). This is generally stated, for the *t*-test, as: 'The difference between the sample means for [insert the two variables] is simply due to random factors, and is not significant', or 'There is no significant difference between the sample means for [insert the two variables]'.

> **Tip** You must not state that there is 'no difference'. Of course there is a difference — sample means vary. What you must state is that any difference is not significant or is due to chance.

Having carried out a *t*-test, you have your calculated *t*-value. You also know the degrees of freedom for comparing the two samples ($n_1 + n_2 - 2$). You are now in a position to use a *t*-table to determine the probability of the null hypothesis being true. Look across the row for the relevant degrees of freedom (go to the next lower value if the exact value is not included in the table) and find where your calculated *t*-value fits between the *t*-values in the table. Now look up and read off the two *p* values at the top. These are the *p* values within which the probability of the null hypothesis being true lies. A value of $p < 0.05$ means that there is a probability of less than five times in a hundred of the null hypothesis being true, and so H_0 is rejected and a significant difference is concluded. Obviously there is a chance of being wrong (up to five times in a hundred). However, different significance levels are recognised, and you should always quote them. Table 10 shows the different significance levels.

Table 10 Different significance levels in statistical tests

Probability of H_0 being accepted (p value)	Asterisked p value and outcome of test	Significance level
greater than 0.05	$p > 0.05$, accept H_0	No evidence of significant difference
between 0.05 *and* 0.01	$p < 0.05^*$, reject H_0	Significantly different, 95% level
between 0.01 *and* 0.001	$p < 0.01^{**}$, reject H_0	Highly significantly different, 99% level
less than 0.001	$p < 0.001^{***}$, reject H_0	Very highly significantly different, 99.9% level

The number of asterisks following the p value denotes the level of significance as indicated above.

You must remember that a *t*-test is not an end in itself. It is simply a tool that allows you to make a decision about significant difference. If you do find a significant difference then you must use your biological understanding to suggest an explanation.

> **Tip** You should have practised the use of a calculator to determine statistical parameters. The following procedure is suggested:
> 1 Switch on your calculator, go to statistical mode and enter your data. (Before entering data, you may need to clear the memory of any previous data entered.)
> 2 Determine n — you know this, but this is a check that all the data have been entered.
> 3 Determine $\bar{x}$ — you require this, but this is also a check that the data have been correctly entered.
> 4 Determine $\hat{\sigma}$ – on most calculators this will denoted by σ_{n-1} (though it may be $x\sigma_{n-1}$, s or sx).
> 5 Square the above (to get $\hat{\sigma}^2$), divide by n (to get $\hat{\sigma}_{\bar{x}}^2$) and then root this value to determine $\hat{\sigma}_{\bar{x}}$ — this can be used either in calculating confidence limits or in the *t*-test.

Note that the statistics sheets used in A-level biology are provided on the CCEA biology microsite — at the back of the specification. You should download them. These show the equations for confidence limits and the *t*-test both in terms of $\hat{\sigma}$ and $\hat{\sigma}_{\bar{x}}$. You may be provided with, say, $\hat{\sigma}_{\bar{x}}$ in a question in Unit A2 2. Make sure that you use the appropriate equation.

Statistical analysis of frequencies – the chi-squared (χ^2) test

In some investigations the DV is a count or frequency (that is, a number of items). This could be the number of fruitfly phenotypes in a genetic cross, or the number of woodlice in a choice chamber one side of which is light and the other dark, or the number of earthworms in a series of fields. The numbers counted (referred to as the **observed frequencies**, O) will differ at random from those expected on the basis of a reasoned hypothesis. That hypothesis will depend on the investigation: in a genetic cross it may be a 3:1 ratio; in the earthworm counts in, say, three fields, equal counts in the fields might be expected. This allows you to calculate the **expected frequencies** (E). The χ^2 test allows you to decide whether or not the observed frequencies deviate significantly from those expected. The formula for χ^2 is:

$$\chi^2 = \sum \frac{(O - E)^2}{E}$$

You must remember that *the sum of the observed frequencies must equal the sum of the expected frequencies.* Again, a null hypothesis (H_0) is set up. A generalised one would be: 'Any differences between the observed frequencies and the expected frequencies are due to chance alone and are not significant'. However, H_0 should be stated specifically for the investigation that you are carrying out:

- For a genetic cross — *The numbers of wild-type and vestigial-winged flies only differ from those expected on the basis of a 3:1 ratio as a result of random factors.*
- For the numbers of earthworms in three fields — *The numbers of earthworms in the three fields are equal and any deviation of the observed counts from this is not significant.*

You use a χ^2 table to determine the probability of this H_0 being true. Look across the row for the relevant degrees of freedom, $n - 1$ (where n is the number of categories), find where your calculated χ^2 value fits between neighbouring tabular χ^2 values; look up and record the two corresponding p values. These are the p values within which the probability of H_0 being true lies.

> **Tip** When quoting p values always ensure that the chevrons are the right way round. Check by ensuring that the larger number is shown as being greater than the lesser number, e.g. $0.05 > p > 0.01$.

Note that the chi-squared test can only be used with raw data (the counts); it cannot be used with processed data such as means or percentages.

Planning for statistical analysis

You must clearly state what measurements (or counts) you are going to take. This is the DV. You should then show the format by which you are going to record results. This will be in the form of a table. Remember that the IV is shown in the first column and the measurements for the replicates and means of the DV are presented in subsequent columns (see page 46).

You must then decide on the most appropriate method of statistical analysis for your investigation. You must present the analysis as part of your planning – it is not sufficient simply to generate results and then ask which method you should use. The diagram in Figure 36 should help you with this aspect of planning. You need to think about whether you are comparing sample means or frequency data (in which case you use a χ^2 value); if you are comparing sample means, you use the t-test for comparing two sample means and confidence limits for more than two sample means.

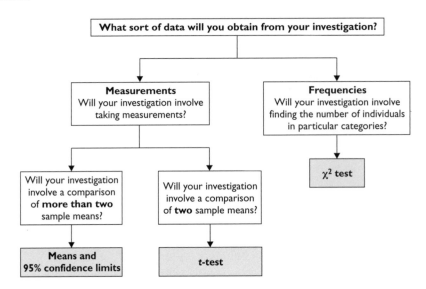

Figure 36 Selecting the appropriate method of statistical analysis

Once you have decided on the appropriate method of statistical analysis you will be in a better position to determine how much replication is required. Obviously, the more replication the better. However — and you must discuss this in your write-up — you have to consider the amount of work you will have to undertake in order to take a single measurement. If it is too time consuming then your replication must, by necessity, be limited. You will also have to consider how well variables can be controlled. If they can be well controlled, as in, say, enzyme experiments in the laboratory, then a minimum of three replicates might suffice though five would be better. However, if you are undertaking an ecological investigation, where variables cannot readily be controlled, then a replication of 30 might be required. Generally, when taking measurements a replication above 30 does not increase the reliability of the data in any substantial way.

If you are working with frequency data, for example in determining the number of seeds germinated or not germinated, then the number of counts taken needs to be large. This is partly because it takes little effort to decide whether or not, for example, a seed has germinated. More importantly, large numbers of data are required for the successful application of the χ^2 test. So, if you are making counts (using frequency data) then you need to have hundreds of data. In a seed germination experiment, for example, you would need to be using around 200 seeds in any one set-up.

Implementation

For details of this skill area, see pages 37–42.

Recording data in a table

For details of this skill area, see pages 43–47.

Statistical analysis

What you should present will depend on the method of statistical analysis that you have planned for. You should revise the appropriate method (see pages 79–83).

Plotting means and 95% confidence limits

There should be a table of statistical parameters — this is in addition to your table recording the data. Table 11 shows the statistical parameters for an investigation into the effect of pH on fungal amylase.

Table 11 The statistical parameters for an investigation into the effect of pH on fungal amylase — the sample size (n) is 5 and, for 95% confidence limits, t = 2.776

pH	Mean time taken for starch digestion $(\bar{x})$/s	$\hat{\sigma}$	$\hat{\sigma}_{\bar{x}}$	95% confidence limits
4.4	780	95.0	42.5	780 ± 118 (662 to 898)
5.1	630	61.3	27.4	630 ± 76 (554 to 706)
5.6	380	57.9	25.9	380 ± 72 (308 to 452)
6.0	92	14.5	6.5	92 ± 18 (74 to 110)
6.4	260	52.3	23.4	260 ± 65 (195 to 325)
7.2	580	48.3	21.6	580 ± 60 (520 to 640)
7.6	725	84.5	37.8	725 ± 105 (620 to 830)

The 95% confidence limits are also shown in Table 11. Figure 37 shows a graph in which means and 95% confidence limits for fungal amylase activity are plotted against pH.

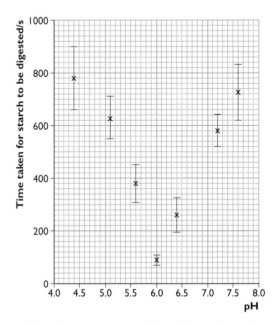

Figure 37 The effect of pH on the activity of fungal amylase — mean time taken for starch digestion and 95% confidence limits plotted

Note that the graph must have a caption, have labels for each axis, be appropriately scaled and, of course, have correct plotting of means and confidence limits. For example, scaling would be inappropriate if the x-axis had started at 0, or if any of the intervals was not constant (0.5 in this case). Also note that the pH values used do not need to be whole numbers, but they do need to be plotted accurately.

Using the *t*-test

There should be a table of statistical parameters. This should show the standard deviation of the mean and 95% confidence limits, even though you are undertaking a *t*-test. The reason for this is that you will need (in the 'interpretation' section) to assess and comment on the reliability of the data. (It is possible to assess reliability from the standard deviation (error) of the mean.) Table 12 shows statistical parameters for an investigation into the effect of bile salts on the action of lipase: the time taken for lipids to be digested was measured and converted into a rate by calculating the inverse; in one set-up, the treatment, a solution of bile salts was added, while in the control distilled water was added in its place; the number of replicates in each case was 10. Note that it is essential in this circumstance to calculate statistical parameters after processing the raw data, that is, converting the 'times' to 'rates' before calculating mean etc.

Table 12 The statistical parameters for an investigation into the effect of bile salts on the action of lipase

	Rate of lipase activity/$10^{-3}\,s^{-1}$	
	Treated — solution of bile salts added	Control — distilled water added (in place of bile salts)
$\bar{x}$	5.32	4.07
$\hat{\sigma}$	0.811	0.909
$\hat{\sigma}_{\bar{x}}$	0.256	0.287
95% confidence limits	5.32 ± 0.58 (4.74 to 5.90)	4.07 ± 0.65 (3.42 to 4.72)

The means and confidence limits can also be plotted on a bar chart to help in assessing the reliability — see Figure 38.

Even though 95% confidence limits may be compared, the appropriate — and stronger — statistical test is the t-test. For this you will need to show your calculation. For the investigation described:

H_0: *The activity of lipase treated with a solution of bile salts (measured as the treatment mean) is not significantly different from its activity with distilled water added (the control mean).*

Calculated $t = 3.247$

Sample size: $n_1 = 10$, $n_2 = 10$

Degrees of freedom $= 18$

$0.01 > p > 0.002$**

Since p is less than 0.05, the null hypothesis is rejected. There is a highly significant difference (at the 99% level) between the treatment and control means.

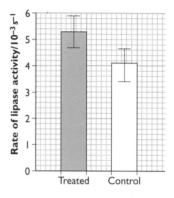

Figure 38 A bar chart comparing the effect of bile salts on the activity of lipase; the treatment included a solution of bile salts and in the control this was replaced with distilled water

Looking at Figure 38 we can see that the activity of lipase is improved by the addition of bile salts. In your 'interpretation', you then enter into a discussion as to why this might be the case.

> **Tip** The rigour of the t-test can be seen in the statistical analysis above. A comparison of the confidence limits suggests that there is (just) no significant difference between the means — see Figure 38. However, the more appropriate t-test concludes a highly significant difference between the means.

Using the chi-squared (χ^2) test

In an investigation into habitat preferences of woodlice, 20 woodlice are released into a choice chamber, one half of which is open to the light while the other side is darkened. After 20 minutes in the chamber, counts are taken of 8 on the light side and 12 on the dark side. The experiment is repeated 10 times (with different woodlice) and the results summed; in total 80 are found on the light side, 120 on the dark side. The results for 200 woodlice should be more reliable than for 20.

In this situation, the χ^2 test is applicable. The test should start with a null hypothesis:

> Woodlice show no preference for light or dark conditions, any deviation from the expected value being due to chance alone.

The expected values are calculated — if there is no preference for either condition, and a total of 200 woodlice were released, then each side would be expected to have 100 woodlice. The calculation of χ^2 is shown below:

Categories	Observed (O)	Expected (E)	(O – E)	(O – E)^2	$\frac{(O-E)^2}{E}$
Light side	80	100	–20	400	4
Dark side	120	100	20	400	4

$\chi^2 = 8.0$
Degrees of freedom = 1 (since there are only 2 categories)
$0.01 > p > 0.001**$

Since p is less than 0.05, the null hypothesis is rejected. There is a highly significant (p is less than 0.01) deviation from the expected values. In this investigation, woodlice have shown a preference for dark conditions.

Note that if only the first set-up had been used (8 in the light side and 12 in the dark) the χ^2 value would have been 0.8, and $0.5 > p > 0.1$, so H_0 would have been accepted (there is no significant difference).

Interpretation

In your interpretation at A2, you must refer to the results of statistical analysis. You should assess and discuss the reliability of the data, using statistical evidence.

For measured data, where you have calculated sample means, you should refer to the relative size of the standard deviation of the mean ($\hat{\sigma}_{\bar{x}}$). If this is small relative to the size of the sample mean then the data are regarded as being reliable. However, if $\hat{\sigma}_{\bar{x}}$ is relatively large then you would be less certain about the reliability of the data, and so conclusions involving such data would need to be tentative.

For frequency data, you must emphasise the importance of large numbers of counts. In the experiment described above you saw that a count of 80 woodlice in the light and 120 in the dark resulted in a highly significant difference; while a similar ratio with only 20 woodlice used (8 in light, 12 in dark) indicated no significant difference. It is not sufficient just to say that you used large numbers. You must use your data to explain why large numbers provide greater reliability.

Having assessed and commented on reliability, you must then try to identify any trends or differences that might be evident in your data. You should again make use of the statistical evidence when:

- comparing means and 95% confidence limits
- referring to the results of a t-test
- referring to the results of a χ^2 test

Evidence for statistical differences means that you have a firm foundation on which to base further discussion. Remember to quote the level of significant difference, if appropriate. Then, using your biological understanding you should attempt to provide reasons for any trends or differences identified. Your explanation must be concise — irrelevancies will take away from your reasoning — and yet be appropriately detailed. You must give the impression that you have a thorough understanding of the topic.

If you find that a null hypothesis has to be accepted you still have much to discuss. Is the level of significance just short of the critical level (e.g. $p < 0.1$)? Would increasing the amount of replication have improved the situation? Even so, you still should be able to discuss the original hypothesis and suggest improvements in your experimental design and lines for further investigation.

You should read through the section on 'interpretation' at AS again — see pages 54–55.

Evaluation

The section on 'evaluation' at AS is relevant to your practical investigation at A2. You should read through this again — see pages 56–59.

However, you should in addition consider the following:

- The appropriateness of the range of the IV used in your experimental design; or, if you had designed treatment and control experiments, the appropriateness of the conditions used for the experimental treatment. With respect to the former, if you had investigated the effect of pH on the activity of fungal amylase, you might wish to undertake a narrower range of pH buffers in order to obtain a better estimate of the optimum pH for the activity of this enzyme. With respect to treatment and control experiments, if you had investigated the effect of bile salts on lipase activity, you might have improved the design if you had added the bile salts to the lipids prior to adding the lipase.
- How another IV could have been investigated. You could consider what other investigations might be carried out to provide more information on which to base your conclusions. For example, if you had been investigating the effect of light intensity on the production of gas (assumed to be oxygen), you could investigate the effect of temperature on the rate of photosynthesis. In this way you could see if the heating effect of the lamp did influence the results.

investigational and practical skills in A2 Biology

Assessment of an investigation at A2: exemplars and comments

At A2, you will be assessed in a *single investigation*. This may involve any topic at A-level (including those at AS) as long as it allows for the assessment of the appropriate skill areas.

In the investigation you will be assessed over the following **skill areas**:
- Developing a hypothesis
- Planning a procedure
- Planning for analysis
- Implementing
- Tabulation of raw and derived data
- Statistical analysis
- Interpretation
- Evaluation of the experimental (practical) procedure

Each of these areas has **five skills** (with mark descriptors), each of which will be marked by your teacher on a three-point scale: 0, 1 or 2 marks.

0 marks will be awarded under the following circumstances:
- There is no evidence of the skill having been demonstrated, e.g. there is no caption on the graph.
- The attempt to demonstrate the skill is not appropriate, e.g. neither 'my results' nor 'table of results' is an appropriate caption for a graph.
- The attempt to demonstrate the skill lacks evidence, e.g. simply saying, in an evaluation of results, that 'my results are valid' without offering any supporting evidence.
- The attempt to demonstrate the skill offers only an incorrect answer, e.g. 'a high % transmission indicates a lot of pigment present'.
- The attempt to demonstrate a skill lacks relevance, e.g. in an experiment on the effect of pH on enzyme activity, only discussing the effect of temperature.

2 marks are awarded for the demonstration of a high level of competence in demonstrating a skill. **1 mark** is awarded where there is a failure to demonstrate full competence in a skill.

The assessment of the skill areas is illustrated below by the use of two investigations. The first investigation, *The effect of biotin on the growth of a yeast population*, is used to illustrate the assessment of **skill area A1** (Developing a hypothesis), **skill area A2** (Planning a procedure), **skill area A3** (Planning for analysis) and **skill area B1** (Implementing). The second investigation, *Investigating resistance to desiccation of three species of rocky shore seaweeds*, is used to illustrate the assessment of **skill area B2** (Tabulation of data), **skill area C1** (Statistical analysis), **skill area C2** (Interpretation) and **skill area C3** (Evaluation of the experimental procedure).

Investigation 1: the effect of biotin on the growth of a yeast population

Student B's responses in skill areas A1, A2, A3 and B1

Skill area A1 (Developing a hypothesis)

The problem was presented as: 'Does biotin have an effect on the growth of a yeast population?' Biotin is a member of the B-complex of vitamins. Vitamins are vital in the diet and are used in the cell's metabolism. So, the addition of biotin should influence the growth of a yeast population. Yeast is a unicellular microorganism, which is part of the kingdom Fungi. It feeds by extra-cellular digestion, a process involving secretion of enzymes followed by absorption of the digestive products. The culture medium supplied contains glucose and some salts, so there is no need for the release of enzymes since glucose is soluble. My hypothesis is: 'Supplying biotin to yeast will affect the rate that a population of yeast grows'.

✍ Student B scores 6 marks, as follows:

Skill	Examiner's comment	Mark
Outline of research relating to the problem under investigation	The student has undertaken research into both biotin as a vitamin and yeast, and is awarded 2 marks for the outline of research. There are limitations but these are penalised elsewhere.	2
Use of appropriate biological understanding	One such limitation relates to the depth of biological understanding demonstrated (e.g. there is no detail about biotin in an organism's metabolism) and for this only 1 mark is scored.	1
Selection of information relevant to the investigation	Further, there is more about extracellular digestion in yeast than there is about its nutritive requirements, and so, for relevance of information, only 1 mark is scored.	1
Sequencing/linking of ideas in presenting an explanation	While both biotin and yeast are referred to there is no link between the two (as in the role of biotin in yeast metabolism) and so 1 mark is awarded for linking of ideas.	1
Precise statement of a hypothesis in suggesting an appropriate explanation	While the section ends with a clear statement about biotin and yeast population growth, this is not sufficient since the effect is not stated. The statement of a hypothesis is not directional, that is, it does not say whether the effect is to increase or decrease growth, and so 1 mark only is awarded.	1

Skill area A2 (Planning a procedure)

Apparatus and materials include: yeast culture (supplied); culture solution containing glucose and salts (supplied); biotin solution (supplied); water bath; two sterile 250 cm³ conical flasks; sterile distilled water; 1 cm³ syringes; teat pipettes; haemocytometer kit; ethanol; and microscope.

100 cm³ of culture solution (glucose plus salts) is added to two sterile flasks. To one flask 1 cm³ of biotin solution is added. To the other flask 1 cm³ of sterile distilled water is added. This is the control experiment. To each flask 1 cm³ of the yeast culture, which has been shaken beforehand to ensure the yeast cells are evenly dispersed, is added. Different 1 cm³ syringes are used each time. A water bath is set up for 25°C and the flasks put in the water, so that the temperature is kept constant. The flasks are left in the water bath for 24 hours.

After 24 hours, there should be appreciable growth of the yeast population. Cells are then counted using a haemocytometer. Both the haemocytometer and cover glass need to be thoroughly cleaned with ethanol, so that any cells that are adhering to them from previous counts are removed. The cover glass is positioned correctly on the haemocytometer — by breathing on the underside and sliding it horizontally into position. This ensures that there is exactly 0.1 mm between the bottom of the cover glass and the counting surface. Shake the culture flask with added biotin, to distribute cells evenly, and, using a teat pipette, transfer one drop of the yeast culture onto the surface of the haemocytometer so that the solution is drawn under the cover glass by capillary action. The haemocytometer is placed on the microscope stage and adjusted so that the counting grid is in the centre of the field of view at ×400 magnification. The haemocytometer is left for 5 minutes for the yeast cells to settle on the grid. The number of yeast cells in five Type-C squares in each of five Type-B squares (a total of 25 squares) is counted and recorded. This is then repeated for a further four samples from the yeast culture flask (so that there are, in total, five repeats). The whole procedure is repeated for the control yeast culture to which no biotin was added.

ℯ Student B scores 8 marks, as follows:

Skill	Examiner's comment	Mark
Use of appropriate apparatus and materials	The student provides a detailed procedure: the list of materials is comprehensive.	2
Appropriate range for the independent variable (or appropriate control for the experimental treatment)	The experimental treatment and control experiments are appropriate.	2
Procedure presented as an ordered sequence to follow	The procedure was well sequenced.	2
Appropriate measures to control variable — fair test	There were appropriate measures taken to control other variables, such as volumes, temperature and representative samples from the yeast culture. (Note that, while there is no explicit reference to a standard technique for dealing with cells touching the sides of the counting squares, understanding of this is demonstrated in the report for the next skill area.)	2

Skill	Examiner's comment	Mark
Prediction of the outcome of the experiment with reference to the measurement of the dependent variable	The student has failed to make a prediction of the outcome of the experiment with reference to a comparison of yeast cell counts and so, for this skill, is awarded no marks. An appropriate prediction would be: 'the cell counts in the yeast culture with biotin added will be significantly higher than the cell counts in the control yeast culture in which biotin was absent'.	0

Skill area A3 (Planning for analysis)

The number of yeast cells lying within the square, including those touching the top and left sides of squares but not those touching the right and bottom sides, in five Type-C squares in each of five Type-B squares, are recorded. This is done for both the culture with biotin added and the control culture. For each a table for recording results would be arranged as shown below:

Type-B squares	Number of yeast cells in 25 Type-C squares				
	Type-C squares				
	1	2	3	4	5
1					
2					
3					
4					
5					

Similar tables are constructed for each of the five repeats for both the biotin-added culture and the control. From the total cell counts, the means for the two cultures are calculated, along with the standard deviation and the standard deviation (error) of the mean. These are shown in an additional table:

		Total cell counts for 25 Type-C squares	
		Culture with biotin	Culture without biotin
Replicates	1		
	2		
	3		
	4		
	5		
Mean ($\bar{x}$)			
Standard deviation ($\hat{\sigma}$)			
Standard deviation of the mean ($\hat{\sigma}_{\bar{x}}$)			

The most appropriate method for statistically comparing the results is via a *t*-test.

The minimum replication is three. I am suggesting a replication of five, since a higher number of repeats should improve the reliability of the results. Even with five repeats there is a lot of work in taking cell counts with a haemocytometer. For example, this would mean a total of 125 Type-C squares being counted, for each of the treatment and control cultures. This amount of work is only feasible from a 'time' point of view by using the results of other students (who adopt the same planned procedure). Taking repeats also allows anomalous results to be spotted.

Student B scores 8 marks, as follows:

Skill	Examiner's comment	Mark
Clear statement of the measurements of the dependent variable to be recorded	The student has failed to show how a calculation of cell density is going to be carried out, i.e. the number of cells per mm^3 of culture taking account of the dimensions of a Type-C square and the 0.1 mm depth of culture in the haemocytometer. So, for 'measuring cell density' only 1 mark is awarded.	1
Format for the table in which results are to be recorded	The student has made appropriate planning for the table format (with a clear indication of what is to be recorded).	2
Selection of the most appropriate method of statistical analysis and justification of this	The student has correctly chosen a t-test for statistical analysis, but has failed to provide a justification as to why this is the most appropriate technique, and so only 1 mark is awarded. The student should have noted that two sample means were being compared.	1
Decision about the appropriate amount of replication required	The student has decided on an appropriate amount of replication and its impact upon reliability.	2
Consideration of time and material constraints in deciding the appropriateness of the level of replication	The constraints on further replication have been appropriately considered.	2

Skill area B1 (Implementing)

The implementation of the planned procedure was observed and marked by the teacher.

Student B scores 10 marks, as follows:

Skill	Examiner's comment	Mark
Skilful handling of apparatus and materials	Setting up of the yeast culture, taking samples and use of the haemocytometer were all well carried out.	2
Adherence to appropriate safety procedures, etc.	The student wore a lab coat while doing practical work and was able to take samples of the yeast culture without spillages.	2

Skill	Examiner's comment	Mark
Organised and methodical carrying out of procedure	The student worked carefully through the procedure.	2
Adhering to sequence of planned actions and making modifications as appropriate	All actions in carrying out procedures were well sequenced.	2
Measurements at the level of precision required	The student was able to measure at the level of precision required and, in particular, followed rules for dealing with cells at Type-C square boundaries.	2

Investigation 2: resistance to desiccation of three species of rocky shore seaweed

Different seaweed species are found in different locations of a rocky shore. *Pelvetia canaliculata* is found on the upper shore, *Ascophyllum nodosum* on the middle shore while *Fucus serratus* is confined to the lower shore area. Specimens of these seaweeds were removed from a shore, immersed in seawater and brought back to the laboratory. Twenty specimens of each species were trimmed so that they were all approximately 100 g in mass, surface dried on paper towels and the initial mass of each recorded. They were then placed on a bench in the lab and exposed to the air for 2 hours. After this time, the final mass of each was recorded and the % loss in mass calculated.

Student B's responses in skill areas B2, C1, C2 and C3

Skill area B2 (Tabulation of raw and processed (derived) data)

The student presented raw data (initial and final mass of seaweeds) in a table placed in an appendix to the report. Assessment is of the tabulation of the processed (derived) data — except for assessing inclusion of 'units of measurement'.

Table of results for % loss in mass for three species of seaweed

Seaweed species	Replicates	Mean
Pelvetia canaliculata	9.6*, 4.4*, 4.3*, 7.3*, 5.0, 3.2, 4.9, 4.8, 4.7, 12.5, 8.0, 2.4, 5.1, 6.2, 2.8, 5.0, 4.5, 1.5, 2.1, 2.2	5.03
Ascophyllum nodosum	5.1*, 3.6*, 4.6*, 8.1*, 5.1, 5.7, 3.4, 6.7, 2.2, 5.2, 11.2, 4.2, 4.4, 4.7, 7.7, 14.0, 6.0, 2.5, 4.3, 4.4	5.66
Fucus serratus	18.0*, 12.3*, 10.4*, 10.7*, 14.9, 13.5, 13.1, 9.7, 12.5, 12.9, 14.3, 5.9, 13.6, 12.5, 14.4, 20.0, 8.6, 5.6, 13.2, 13.3	12.47

*Student's own results — these were presented in a separate table for assessment and included the same weaknesses as illustrated in this table (while replicate results are presented here to allow for the assessment of variation within the replicates).

Student B scores 8 marks, as follows:

Skill	Examiner's comment	Mark
Organisation of the table with the independent variable in the first column or row and the dependent variable in the body of the table	The independent variable here is the species of seaweed. The student has this in the first column, which is appropriate.	2
Caption including the independent and dependent variables and the material used	However, the caption for the student's table, while noting the variables, does not indicate that mass loss is due to '2 hours of exposure to air' (it might be due to osmotic loss) and so only 1 mark is awarded.	1
Logical construction of table — to facilitate analysis	The table has a structure that facilitates analysis (e.g. the replicates can be readily compared to identify any anomalies and/or assess the variation within the replicates).	2
Explanatory column headings	The student scores 1 mark for the species of seaweed but fails to gain the second mark since 'replicates' and 'mean' are not sufficient headings. They should have been '% loss in mass among the replicates' and 'Mean % loss in mass' respectively to gain the second mark.	1
Units of measurement for all variables and calculated values	The table of raw data (not shown here) indicated that measurements were made in grams and this was included in the column headings for initial mass and final mass.	2

Skill area C1 (Statistical analysis)

Table of statistical values for % loss in mass for three species of seaweed

Statistic	*Pelvetia canaliculata*	*Ascophyllum nodosum*	*Fucus serratus*
$\bar{x}$	5.03	5.66	12.47
$\hat{\sigma}$	2.69	2.84	3.44
n	20	20	20
$\hat{\sigma}_{\bar{x}}$	0.60	0.63	0.77
Degrees of freedom	19	19	19
t-value from table for $p = 0.05$ and at 19 d.f.	2.093	2.093	2.093
Upper 95% confidence limit	$5.03 + (2.093 \times 0.60)$ $= 6.29$	$5.66 + (2.093 \times 0.63)$ $= 6.98$	$12.47 + (2.093 \times 0.77)$ $= 14.08$
Lower 95% confidence limit	$5.03 - (2.093 \times 0.60)$ $= 3.77$	$5.66 - (2.093 \times 0.63)$ $= 4.34$	$12.47 - (2.093 \times 0.77)$ $= 10.86$

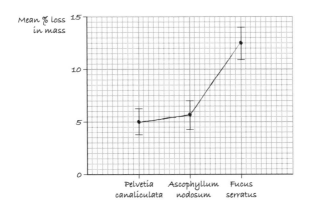

The mean % loss in mass for three species of seaweed

2 Student B scores 8 marks, as follows:

Skill	Examiner's comment	Mark
Table of statistical parameters	The student has presented a comprehensive table of statistical parameters (including the mean, standard deviation of the mean, *t*-value for calculating confidence limits and the 95% confidence limits).	2
Appropriate null hypothesis (*t*-test or χ^2 test) or accurate calculation of confidence limits	The 95% confidence limits have been accurately calculated.	2
Accurate calculation of *t* or χ^2 value or appropriate caption for graph of means and confidence limits	A graph plotting means and confidence limits is presented. However, the caption is incomplete in that there is no reference to the data plotted as means and 95% confidence limits — so, for 'appropriate caption', only 1 mark is awarded.	1
Correct determination of probability value for test or appropriate scale and axes labels and units	While there is enough here regarding scaling and labels for 1 mark, the second mark is lost because the graph (scaling) is inappropriate. The student has drawn a line graph and joined the plotted points when the seaweed species are obviously discrete — there is nothing lying between the different species!	1
Correct decision regarding the null hypothesis or correct plotting of means and confidence limits	Besides the limitations shown above, the 'means and confidence limits' are accurately plotted.	2

Skill area C2 (Interpretation)

To assess the reliability of the results I looked at the width of the 95% confidence limits compared with the mean of each species of seaweed. The confidence limits are moderately large, which shows that it is possible that the results are unreliable. The results may include some anomalies. For example, with *Pelvetia canaliculata*, while most values lie between 4 and 6 (mean % mass loss is 5.03) there is one result at 12.5.

Comparing the mean % loss in mass for the three species of seaweed, it is possible to identify trends. *Fucus serratus* has the largest mean % loss in mass at 12.47%. This appears to be a significantly greater loss since the 95% confidence limits do not overlap with the 95% confidence limits for the other two species. However, while *Ascophyllum nodosum* has a higher mean % mass loss than *Pelvetia canaliculata*, the difference does not appear to be significant. This is because there is considerable overlap in their 95% confidence limits.

The significantly greater loss in mass by *Fucus serratus* can be explained by a bigger loss of water due to the 2 hours of exposure to air. *Fucus serratus* is found on the lower shore and may not need to have any adaptations to reduce water loss (rather like a hydrophyte) since it is most often covered by water. *Pelvetia canaliculata* and *Ascophyllum nodosum* lose less mass because they lose less water. They are mid- and high-shore seaweeds and are often exposed on the shore as the tide recedes. They would be expected to have some adaptations to reduce water loss when exposed to the air.

Student B scores 7 marks, as follows:

Skill	Examiner's comment	Mark
Assessment of the reliability of the data using statistical evidence	The student has a reasonable assessment of the reliability of the data (the width of the 95% confidence limits relative to the size of the means is discussed, along with the link between wide confidence limits and unreliability).	2
Explanation of decision in relation to the reliability of the data	The commentary on reliability, however, is limited to the identification of an anomalous result and so only 1 mark is scored here. The student could have suggested a reason for the anomaly; certainly, the student should have commented on the probable improvement in reliability by increasing replication to 30 specimens of each species, and the possibility of doing this without an appreciable increase in the workload.	1
Trend(s) clearly identified using statistical evidence	The trends are adequately identified by the student and good use is made of the statistical evidence.	2
Explanation of the trend(s)	In explaining the trends, there are limitations. While the student is able to link the amount of water loss to the location of the species on the seashore, some possible adaptations might have been suggested. The situation in plants might have been used, and the student touched on this with the reference to hydrophytes. Presumably, like xerophytes, upper- and middle-shore species possess a thicker or more effective waterproofing layer. This cannot be the waxy cuticle of higher plants, but the mucus outer layer on the seaweeds (the seaweeds felt 'slimy') might fulfil a similar role.	1
Use of appropriate biological understanding	The limitation in offering an explanation also means that biological understanding has not been complete.	1

A2 Biology

98

Skill area C3 (Evaluation of the experimental procedure)

The independent variable in this investigation was the species of seaweed. Three different species from the rocky shore was sufficient to provide a measurable trend in the results.

Other than the independent variable and the dependent variables, other variables were controlled. For example, the seaweeds were all exposed to the same atmosphere — the laboratory — for the same period of time — 2 hours. Also, all the specimens were kept in seawater until just before exposure to air.

The dependent variable was measured as the loss in mass of the specimens converted to a percentage loss in mass. Each seaweed specimen weighed approximately 100 grams, so reading the balance to 1 decimal place gave me four significant figures and a high level of accuracy.

All specimens lost mass. I feel confident in assuming that the loss in mass was due to water being lost by evaporation from the seaweed specimens, and so I consider this experiment to be valid.

This experiment could be repeated for other seaweed species such as *Fucus spiralis* and *Fucus vesiculosus*.

🖉 Student B scores 6 marks, as follows:

Skill	Examiner's comment	Mark
Appropriateness of the range of the independent variable or appropriateness of the control and treatment experiments	The student notes that there was a measurable difference between the seaweed species (the independent variable). However, there is no attempt to link this to the position of the seaweed species on the shore and so, for 'range of the independent variable', only 1 mark is awarded.	1
Procedures used for the control of variables not under investigation	The student has listed a number of controlled variables though has omitted to note that the specimens exposed to the air were of similar size, and, in particular, has failed to explain the importance of calculating % loss in mass to take account of the (slight) differences in initial mass — for this, only 1 mark is awarded.	1
Appropriateness of measure of the dependent variable	The appropriateness of the measure of the dependent variable — mass loss converted to % mass loss — and the understanding of accuracy is sufficient to be awarded 2 marks.	2
Assessment of the validity of the experimental procedure	The student understands that the investigation relies on the assumption that mass loss is due to the loss of water by evaporation but has not presented any further discussion and so only 1 mark is scored for assessing validity.	1
Outline for further investigation leading on from the present investigation	In putting forward an outline for further investigation, the student has suggested the inclusion of other seaweed species but this is only sufficient for 1 mark. The location of the other seaweed species on the rocky shore should have been included for the second mark.	1

investigational and practical skills in A2 Biology

Answers

Practical Skills in AS Biology

1   Of the four biochemical tests, only Benedict's test requires heat.

2   Since the amino acid was spotted 30 mm from the line then it travelled 84 − 30 = 54 mm; the solvent front travelled 150 − 30 =120 mm from the spotting line; 54 mm divided by 120 mm gives an R_f of 0.45.

3   The lipase is left in the water bath for 10 minutes before adding it to the milk (substrate) so that it is at the treatment temperature when the reaction starts.

4   Controlled variables include: the volumes and concentrations of the starch and amylase; and the temperature at which the reaction occurs. Timing issues relating to the taking of test samples are discussed in the text.

5   Heat will cause the oxygen released to increase in volume, and will also speed up the rate of reaction.

6   The smear of oil will prevent the protease enzyme making contact with and digesting the gelatin.

7   The products of sucrose digestion are glucose and fructose. Glucose and fructose are reducing sugars for which Benedict's test is appropriate.

8   Blue solutions most effectively absorb red light provided by a red filter. This allows a greater range of colorimeter readings to be produced.

9   The immersing sucrose solution is open and not under pressure, so $\psi_p = 0$ and $\psi = \psi_s$.

10 Different cells have different amounts of solutes and those with more solutes will have a lower (more negative) solute potential. This will influence the cells' water potential and, in consequence, may lead to the movement of water between cells.

11 Potassium hydroxide treatment takes place first to analyse the relative amount of CO_2 present, since potassium pyrogallate absorbs both CO_2 and O_2.

12 The 'wicks' in the respirometer tubes are used to increase the surface area for the effective absorption of CO_2.

13 The CO_2 taken up by the aquatic plant should cause the pH to increase, since CO_2 in solution is acidic.

14 In 'dark' conditions only respiration is taking place and the CO_2 produced causes the indicator solution to turn yellow; in 'light' conditions photosynthesis is taking place more rapidly than respiration and the net uptake of CO_2 causes the indicator solution to turn red.

15 A leaf could be placed on graph paper, its outline drawn in pencil and number of squares covered counted, taking account of those only partially covered. The graph paper squares will be either 1 mm × 1 mm (1 mm^2) or 2 mm × 2 mm (4 mm^2), which is more often the case. The procedure is carried out for all leaves and the areas summed to provide an estimate of the surface area of the leaves of the shoot.

16 The most appropriate procedure would be a belt transect. It allows the gradation in the vegetation to be determined and provides a measure of abundance at each sampling point. Quadrats (1 m × 1 m) are laid down along a line of transect down the slope. Approximately 20 quadrat positions is advisable so that, if the slope was 100 metres long, the quadrats would be positioned every 5 metres (interrupted belt transect). Different plants should be identified and their abundance determined at each quadrat position.

17 (a) The percentage cover of the entire lawn is 2%. (b) 2% of the lawn area (9 m × 20 m = 180 m^2) is covered, so 3.6 m^2.

18 The overall coverage appears to equate to six small squares. Since each small square represents 4% (1/25), the percentage coverage of creeping buttercup in the quadrat is 6 × 4% = 24%.

Investigational and Practical Skills in A2 Biology

1 8.15 × 4000 = 32 600 cells mm^{-3}.

2 The second sample size is 60 (48 + 12) and so the population size is estimated as 50 × 60 ÷ 12 = 250.

3 This *RQ* value is closest to that for protein but since protein is only a major respiratory substrate during starvation it is unlikely to explain the value of 0.9 obtained here. It is more likely that aerobic respiration is taking place using a mixture of lipid and carbohydrate as respiratory substrate.

4 Because the blue colour at the surface is due to oxygen diffusing into solution from the atmosphere.

5 Wavelength (colour) of light: Use a range of colour filters, for example a blue filter will allow only blue light through. A light meter is placed beside the plant and used to ensure that the light intensity is the same for all filters, since different filters will have a different effect on the light intensity as well as the wavelength. Temperature: Use a series of water baths. Allowance must be made for changes in the solubility of oxygen in water.
Carbon dioxide: Use different concentrations of potassium hydrogen carbonate in which to immerse the plant. Other factors should be kept constant, e.g. a lamp kept at the same distance from the plant.

6 Different plant pigments absorb slightly different wavelengths of light (which is why the pigments have slightly different colours). This means that the range of wavelengths absorbed is increased and so increases the rate of photosynthesis.

7 Because the cells will also contain mitochondria which produce hydrogen (from the dehydrogenase activity in respiration); this will also decolourise DCPIP.